RAINBOW MODULAR SCIENCE & TECHNOLOGY

ELECTRIC CIRCUITS · MAGNETISM
ELECTROMAGNETISM
DOMESTIC ELECTRICITY

J.A. Thomas, M.A. (Educ.)

Head of Science, Castledown School, Ludgershall

Schofield & Sims Ltd   Huddersfield

0 7217 3588 6   Pupil's Book

0 7217 3595 9   Pupil's Book with Teacher's Manual

First printed 1988

Designed by Geoff Bucktrout.   Printed in England by Henry Garnett & Co Ltd, Rotherham.

NOTE TO THE TEACHER

Rainbow Modular Science & Technology aims to provide a self-contained course of study for GCSE, suitable for use with pupils of most abilities.

Traditionally, teachers expend a great deal of energy and time getting together reading materials and questions, devising experiments, and making up worksheets. The modules in this series are an attempt to lift the heavy burden of preparation, by providing ready-made lesson plans in the form of self-sufficient instructional material. The idea is to free the teacher to concentrate on what is most important for GCSE – working with the pupils.

The modular approach offers tremendous economy. Its flexibility – in contrast to the rigidity of the traditional single textbook, which may be in short supply – enables a large number of classes to make use of one set of materials. In addition, CDT departments and those following modular science syllabuses will require only those modules relevant to their needs. When there is a content overlap, the materials may be shared between departments.

The units are carefully structured, offer wide coverage and are supported in each case by a comprehensive *Teacher's Manual*. It is hoped that the series will prove attractive and useful to the non-specialist who may be called upon to help out, as well as to science and physics departments and teachers of CDT.

INTRODUCTION

In the mysterious writings of the ancient prophet Daniel you can read his prediction of an amazing new age that would arrive far into the future, when 'many will travel to and fro, and knowledge will vastly increase'. Daniel wrote these words at a time when travel was rare and the camel was the fastest means of transport, when oxen ploughed the fields, and men laboured from dawn to dusk to make a living, over two thousand years ago, in the mighty kingdom of Babylon.

We live in that age foretold by Daniel, the age of technology, electronics, and nuclear power ... an age when knowledge has indeed exploded, and there are more scientists and engineers alive than have ever lived in all of history put together ... when many travel to and fro in fast cars and supersonic aircraft ... when world travel and communication have shrunk the world into a 'global village' ... and we possess the awesome power to destroy all life on our planet and leave it as a blackened ball, silently spinning in space ...

This book is designed to help you understand the fundamental laws of science that have made such amazing progress possible, and how our technology exploits these laws to improve the quality and comfort of our lives.

The reading sections explain the important ideas. Study them carefully, reading them over two or three times if necessary, and make sure you understand them. If anything is not clear, make a note of it and discuss it later with your teacher. Pay special attention to the sections printed in italics.

Nobody else can learn for you, just as nobody else can build your muscles or make you fit. So get fully involved in the practical ACTIVITIES and develop your practical skills. Do not let yourself become a mere spectator. The ANALYSIS items will test your understanding of the practical work.

In the QUESTION sections, the first items will test your recall and basic understanding of the ideas studied. Usually this will lead to analysis of useful devices or everyday applications of science. The starred * questions are intended to be more challenging than the others: ask your teacher if you get stuck.

Educated people know how to use resource books to dig out information for themselves. For this reason you will find LIBRARY items that give you the chance to develop your library research skills. Your teacher may then give you the chance to develop your speaking skills as well by reporting your findings back to the class.

The greatest resource of any nation is not coal, oil, or even gold deposits. It is the creative power of its people. And creativity can be developed by practice. For this reason there are two other types of question:

INVESTIGATION items require you to design your own experiments that may then be carried out in the lab. DESIGN items require you to solve a problem by inventing and designing.

CONTENTS

1 STATIC ELECTRICITY

Without electricity there could be no TV, radio, computer, record player, power pack or even electric light. Electricity along with energy from oil and coal has revolutionised the life-style of mankind. Yet for thousands of years people lived in ignorance of the electricity that was all around them, and in fact present in every single atom of the human body.

The ancient Greeks noticed the simple electrical effect that amber* beads rubbed on dry cloth would attract bits of fluff, just as a comb rubbed on dry hair will attract bits of paper. But it was over a thousand years before any real progress was made. In 1600 William Gilbert, physician to Queen Elizabeth I, began to study static electricity and the word 'electricity' was invented to describe it. *Static electricity is not moving, in contrast to the current or moving electricity that flows along wires in our modern appliances.*

About 1740 a French scientist, Dufay, found that there were in fact two kinds of static electricity. The one obtained by rubbing glass he called 'vitreous' electricity. The other from rubbing resin he called 'resinous'.

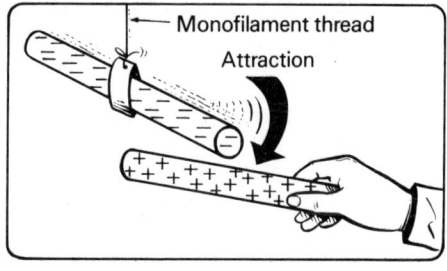

By simple but revolutionary experiments such as that shown in the diagram, Dufay found that vitreous and resinous electricity attracted one another. But he found that resinous repelled resinous, just as vitreous repelled vitreous. Today we would say that *'like' charges repel, and 'unlike' charges attract.*

Later Benjamin Franklin, American diplomat and signatory of the Declaration of Independence, called these two kinds of electricity, positive and negative. He also showed that materials contain equal amounts of each, so that objects are normally neutral (uncharged), which is why we do not notice the vast amount of electric charge all around us and in our own bodies.

*(Latinised) Greek word 'electrum'.

When a positive charge is produced on a glass rod by rubbing with silk, negative charge is rubbed off the glass onto the silk. The two charges are equal so that *each time a positive charge is produced, an equal negative charge must also be produced.*

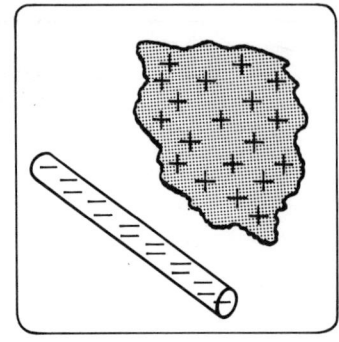

Benjamin Franklin is famous for his kite experiment. He flew a kite in a thunderstorm to prove that lightning was electrical. He found that he could draw a spark by holding his finger near a key dangling from the kite string. Franklin knew that the moist kite string would conduct electricity, so he protected himself by holding the key and string on the end of a length of dry silk ribbon. To keep the ribbon dry, he stood in a doorway. Franklin's idea was proved correct. But the next person to try was struck dead!

We now know that *all atoms contain equal amounts of positive and negative electricity.*

The diagram of a model atom shows the positive protons in the centre (or nucleus) with the same number of negative electrons moving around the nucleus in orbits. It is because they are on the outside that electrons can be rubbed off an atom and can move in wire as current. The positive charges do not move.

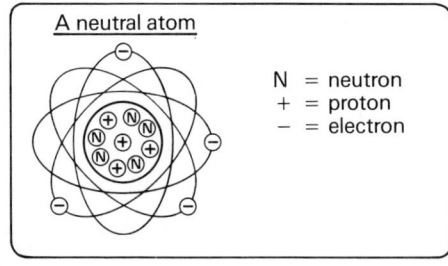

A neutral atom

N = neutron
+ = proton
− = electron

Electrons can move freely through metals, so that metals are good conductors. Materials such as air, plastic and glass are insulators: they do not normally let electrons move.

Electric current is the movement of electrons from negative to positive.

On a dry day, electrostatic experiments work well. On damp days an invisible film of moisture forms a conductive coating and static charges

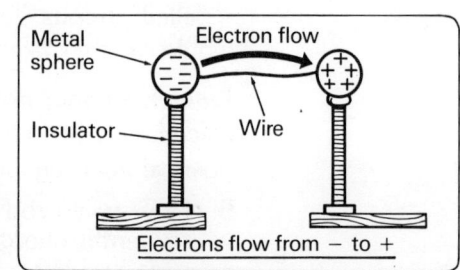

Metal sphere

Electron flow

Insulator

Wire

Electrons flow from − to +

cannot build up. It sometimes helps to dry the apparatus in front of a fan heater or hair drier.

ACTIVITY

a Charge a polythene rod by rubbing with a dry cloth and have it suspended as shown. The rod will have a negative charge which can be used to test other charges.

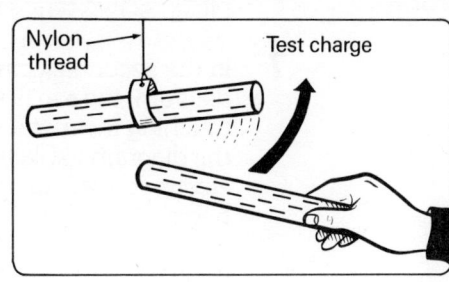

TEST: Any charge that attracts the rod must be positive.
Any charge that repels the rod must be negative.

b Test a variety of materials to find out what kind of charge can be formed on them (conductors will not hold any charge). Then tabulate your results.

Material tested	Rubbed with	Charge: positive/negative/none

c What should you find by testing the cloth you charged the polythene rod with?

Questions

1 What are the modern names for vitreous and resinous electricity?

2 Where does the name electricity come from?

3 Copy the diagram of the model atom. What does it explain about neutral objects?

4 Why will static electricity not stay on metal objects held in the hand?

5 Why are electrostatic experiments best done in dry weather?

6 **Investigation**   An electroscope as shown will spread its leaves after being touched with a charged object, negatively charged in this case.

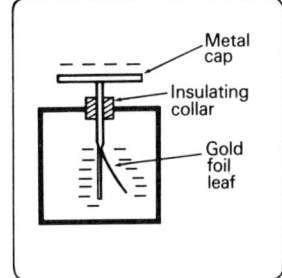

a Why do the electroscope leaves push apart?

b How would you use an electroscope to identify charges as positive or negative (without touching it)?

c How could you test materials to see if they are conductors or insulators?

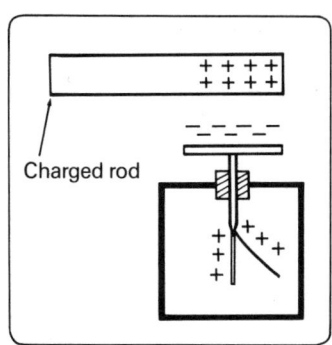

7 In the second electroscope diagram, the rod is not touching the metal. What is the diagram explaining? The electroscope had no charge to begin with.

8 Investigation   You are given a rod with a strong negative charge on it, held on a retort stand. You also have two metal cans plus other simple materials. How could you charge the two cans, one positive, one negative, without touching or using up the charge on the rod? How would you test the cans for charge?

9 Design   In the year 1714, a man called Hawksbee (Isaac Newton's laboratory assistant) built a machine to generate electric charges rapidly. A handle was turned and a glass ball was also used. How do you think it worked? How could one be made with the aid of a bike and some simple laboratory equipment? (Hint: How did you produce static charges in the experiment?)

10 Library   Find out what a van de Graaff generator is. Draw a diagram and explain how it works.

11 Some paint sprayers impart a positive charge to the paint droplets. How does the negative charge on the object being painted speed up the job and also reduce wastage of paint?

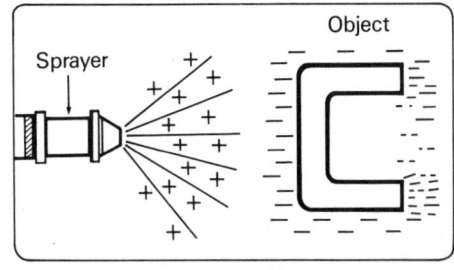

12 A charged rod is held near a metal form as shown. Predict what will happen to the free electrons in the metal. Draw a diagram to illustrate.

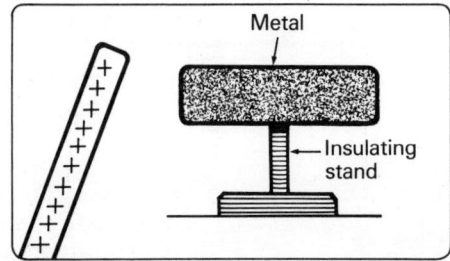

* 13 Early reseachers found that an electroscope would slowly discharge no matter how well it was insulated. The explanation is that there are positive and negatively charged particles (ions) floating about in the air. Explain with the aid of a diagram why a charged object will slowly lose its charge due to the ions.

14 **a** It takes about 2000 volts of electrical potential to provide a spark 1 mm long in air. What does this suggest about the voltage of static charges built up walking across acrylic carpet for example?

b Air liners can have large static charges which may cause sparks and explosions when the metal nozzle of refuelling equipment comes into contact. Suggest a safety procedure to eliminate the danger.

c A table tennis ball hangs between two charged plates as shown. Predict what will happen to the ball and explain why. Why does it work best if the ball is coated with metallic paint or graphite? What factors will control the movement of the ball?

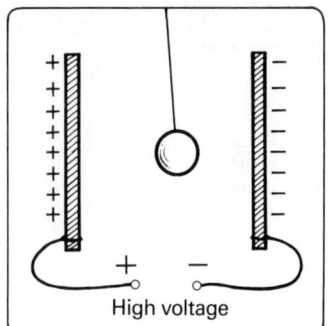

High voltage

2   ELECTRIC CIRCUITS

A circuit is a complete path that returns to its starting point (e.g. a racing circuit). *An electric circuit is similarly a complete path that electrons can move around. When electrons (tiny particles of negative electricity) move through a wire, we say there is an electric current flowing in the wire.* As the electrons circulate, they pick up potential energy from the battery and use it up, usually as heat, as they pass along wires and through lamps. *Materials such as metals, that allow current to pass through, are called conductors. Materials that do not conduct are called insulators.* Metals are all very good conductors. There is a third class of weakly conducting materials called semiconductors. These include carbon (used by Edison in his early light bulbs) and silicon which is used in transistors and microchips.

Conducting materials normally offer some resistance to current. The resistance of a wire depends on various factors such as length, thickness, material, and temperature. These factors may be investigated with the simple circuit shown. In circuit diagrams we assume that the connecting wires have no resistance.

In a lamp, the electrons give energy to the thin metal filament and make it so hot that it glows and gives off light. We say the filament becomes incandescent. The brightness of a lamp is a rough guide to the amount of current in a circuit.

ACTIVITY I   Resistance and Length

a   Set up the circuit shown but with no test wire separating the crocodile clips. The lamp should light to its normal brightness.

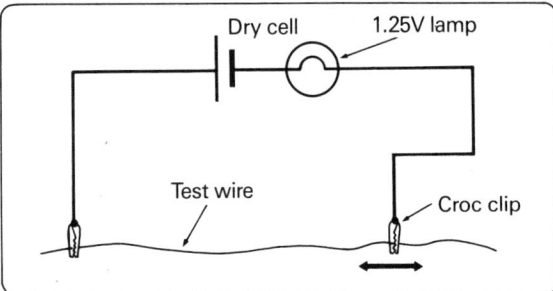

b   Now introduce a length of nichrome wire between the clips. Move one of the clips along the wire to change the length in the circuit. How does the length of wire used in the circuit affect the current (and so the brightness of the lamp)?

c   Copy the circuit and write down your conclusion.

ACTIVITY II   Resistance and Thickness

a   Does a thick wire have more or less resistance than a thin one of the same length? Use your circuit to find out.

b   Write down your conclusion.

ACTIVITY III   Resistances of Different Materials

a   Compare wires made of different metals such as iron, copper, nichrome. For it to be a controlled experiment, any wires you compare must have the same length and thickness.

b   Write down your conclusion.

Questions

1   List the conclusions you drew from the experiment.

2   How could a long and a short wire have the same resistance?

3   What is an electric circuit?

4   We say that air is an insulator, yet sparks are electric currents. Comment on this statement.

5   **Investigation**   The lamp used in the activities has a resistance. How could you find the length of nichrome wire (of an available gauge) that has the same resistance to current as the lamp?

6   When we speak of electric 'current' and say that it 'flows', we are comparing electricity with water. In what ways does water flowing through a hose-pipe behave like electricity flowing through a wire?

7   Suppose you used an electric drill or vacuum cleaner with a cable about 200 metres long. What might happen and why?

8   'Jumper' cables can be used to start an engine with a flat battery by connecting to the battery of another car. There are cheap and expensive types available. The cheap ones give a large voltage drop and are not as effective. Suggest some ways in which the cables might differ.

9   Traditionally the thickness of wire has been given in SWG, Standard Wire Gauge. A common thickness of copper wire for electrical work was 24 gauge. Use the data in the table to plot a graph showing how actual wire diameter in mm is related to SWG number. Comment on the graph.

GAUGE	DIAMETER (mm)
10	3.25
15	1.83
20	0.914
22	0.711
28	0.376
32	0.274
40	0.122

a   From the graph find the diameter of 30 gauge wire.

b   Predict the diameter of 5 gauge wire.

3   ELECTRIC CURRENT

Electric current is measured in amperes (symbol A) and milliamperes (mA). There are 1000 mA in 1A. The instrument used to measure current is the ammeter. For example, a good battery charger has an ammeter to show the charging rate. Fuses are rated in amperes, so that a 13A fuse will blow if the current rises dangerously above that level.

In a simple circuit there is only one path for the current to follow. This is a series circuit. If several lamps and an ammeter were connected in series, the same current would pass through each lamp and the ammeter would show what the value of the current was.

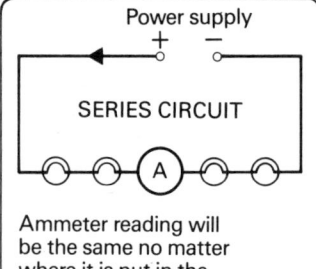

SERIES CIRCUIT

Ammeter reading will be the same no matter where it is put in the circuit

Many circuits are of the branching or parallel type, which means that the current splits up among various paths and then finally reunites just before it returns to the power supply.

Notice the arrows on the circuit diagrams to show the direction of the current. We say that current flows from the positive to the negative of the battery or power supply.

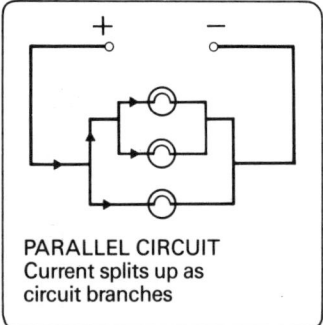

PARALLEL CIRCUIT
Current splits up as circuit branches

Standard circuit symbols are used to draw circuit diagrams. We assume that connecting wires have no resistance. Here is a list of common symbols.

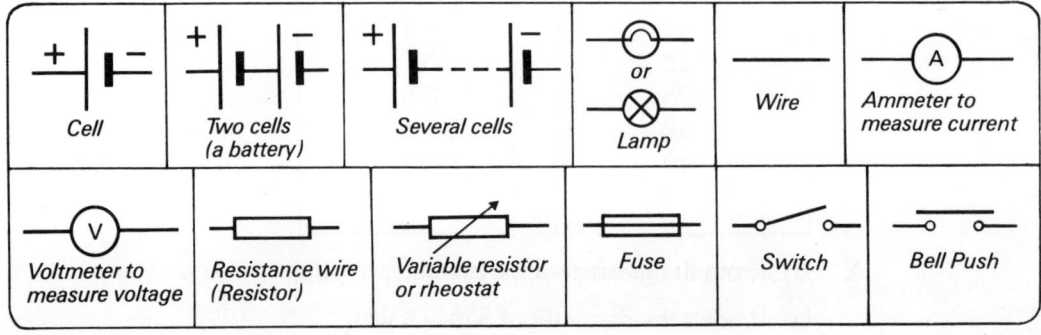

Cell	Two cells (a battery)	Several cells	or Lamp	Wire	Ammeter to measure current
Voltmeter to measure voltage	Resistance wire (Resistor)	Variable resistor or rheostat	Fuse	Switch	Bell Push

In this series circuit, the reading of ammeters A_1 and A_2 will be equal. The current is the same all around the circuit. It does not get 'used up' in some way.

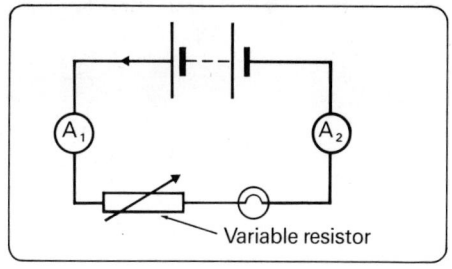

Variable resistor

The variable resistance is used to vary the current in the circuit and so control the brightness of the lamp. When there is more resistance, there is less current.

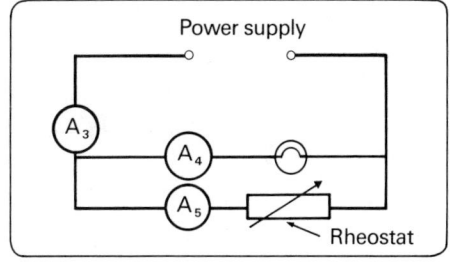

Power supply

Rheostat

In the parallel circuit, the readings of A_4 and A_5 should add up to equal that of A_3. This time the variable resistor (or 'rheostat') varies the current through A_5 and so the lamp should not be affected.

Switches are used to control the current in various parts of a circuit. In the circuit shown Switch A is a master switch. If it is off, both lamps are off. Likewise Switch D is a second master switch and so is not necessary. Note that if the current stops at D, it must stop everywhere else in the circuit too.

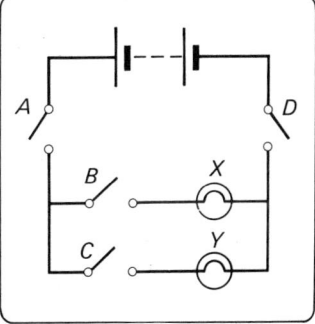

Switch B can only turn lamp X on and off. Switch C controls only lamp Y.

Questions

1   Should ammeters be connected in series or parallel?

2   How many milliamps are there in:

   a 0.5A   b 0.1A   c 0.02A   d 1.0A   e 2.35A?

3 Copy this diagram and tell which lamps each switch can turn on and off.

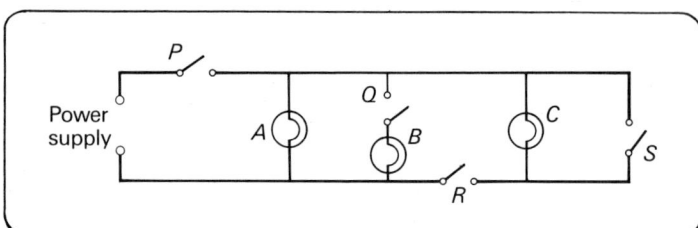

4 Copy the diagrams and tell what would happen when each is switched on, or what is wrong with each one.

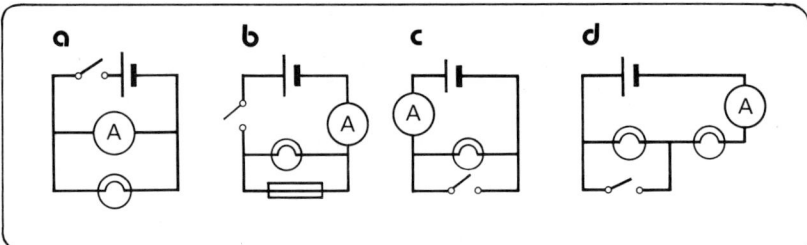

5 The diagram represents a demister heater on a car rear window. If the heater draws a current of 8A, find the currents at the points labelled.

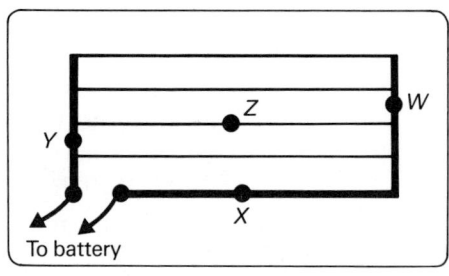

6 In the diagram the two resistors have equal resistance and ammeter A_2 reads 0.5A. Find the readings of the other ammeters.

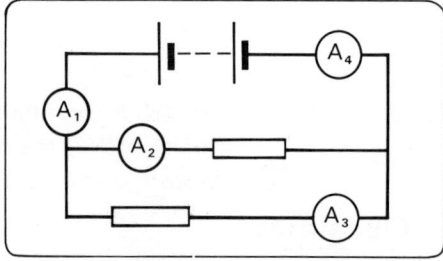

7 On a car, one battery terminal is connected to all the wiring. The other is connected by a thick wire to the metal bodywork. Why is this? How does the circuit work?

8   Are the lights in a house connected in series or parallel? How can you tell? What would happen if the other kind of circuit was used?

9   In the diagram, S_1 and S_2 are two-way (single-pole double-throw) switches. As shown, the lamp is off. Explain how the circuit works and where it is used.

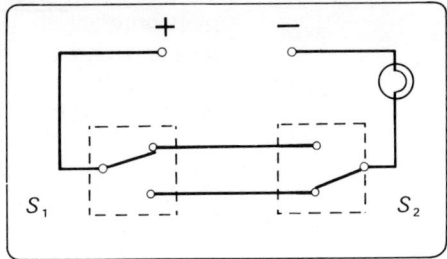

10 A man fitted the following extras to his car using the circuit shown:
spotlamp;
demister heater;
a motor to raise and lower a radio antenna.
Explain what is wrong with the circuit, then re-draw it correctly.

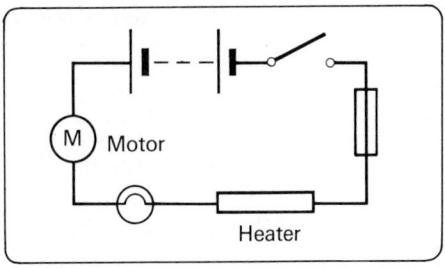

11 **Design**   Draw circuit diagrams for the following.

   a   House door bell, to operate from front and back door.

   b   Safety circuit for guillotine in print works. Motor must only operate when both hands are pulled away from guillotine to operate switches.

   c   Speed control for a slot-car motor.

   d   Stage spotlamps, allowing lamps to be switched independently and controlled for brightness.

   e   A box is fitted with two lamps and a slot. When a strip of metal is inserted in the slot, one lamp comes on. When a strip of card is inserted, the other lamp comes on.

   f   A model to show the principle of a car earthing system, using 3 lamps held in plastic bases, 3 croc leads, plus a strip of mild steel plate measuring about 2 cm by 20 cm. The model is to work from a low-voltage power supply.

12 **Investigation**   Plan experiments to compare systematically the conduction (if any) of electricity by:

   a   distilled water, tap water, salt water

   b   moist sand, gravel, soil.

What variables will be controlled? What measurements will you take? How will you analyse your data?

13 **Investigation** An electric current produces a heating effect in a wire. In a lamp a filament is made so hot that it glows, becomes incandescent. In many domestic appliances current is passed through nichrome wire to produce heat. Examples include immersion heaters, toasters and fan heaters. Plan an investigation to find out how the heat energy output of a coil of nichrome wire varies with the current. Heat energy can be monitored by using it to warm water in a foam plastic cup.

a If you have a 0–5A ammeter, what current values might you try?

b How will you find out what length of resistance wire will allow your power supply to pass the highest current chosen at, say, 6V?

c How will you fit the heating wire into the cup of water without parts of it touching and shorting out?

d What rough trial experiment would you try first to see if a good temperature rise of say 50°C can be produced for your largest current?

e What graph would you plot to analyse your data?

14 **Design** The diagram shows how a mercury switch works. When tilted, the mercury joins the contacts and makes the circuit.

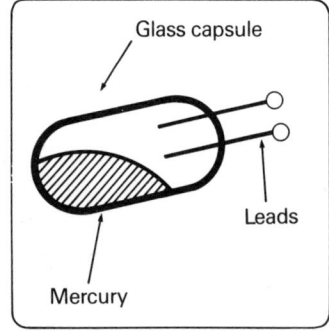

a How could this switch be used to operate a seat-belt warning light in a car?

b How could a mercury switch be designed to stay on when tilted for a fixed time such as 15 seconds?

c How could mercury be used in an electric egg timer?

15 **Design** **a** You are given a 0–1A ammeter and some bare copper wire. How could you adapt the ammeter to read 0–2A?

b Suggest a permanent design for your device so that several could be made and easily fitted to ammeters when needed.

ACTIVITIES

In using an ammeter to investigate circuits, remember that none of your readings will be totally accurate and so the sum of the currents in the branches of a circuit may not quite equal that in the main circuit. If you are using an analogue meter (one with a needle), read it at right angles to the scale to avoid parallax errors. Make sure that the positive terminal of the ammeter is connected to the positive side in the circuit.

SAFETY

Never connect an ammeter alone to a battery as it may be overloaded. Ammeters are only used to measure the current drawn by a device such as a lamp or resistor, never alone. Before switching on, trace the circuit path from positive of the supply to the negative to make sure it is as shown. Have supply on lowest setting to start with.

ACTIVITY I   Series Circuit

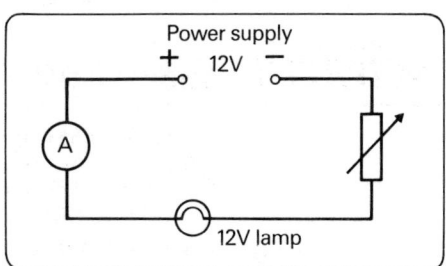

a Connect the circuit shown. Find the current needed to light the lamp:
   dull red, just visible;
   to normal brightness.

b Copy the circuit and record your results.

ACTIVITY II   Parallel Circuit

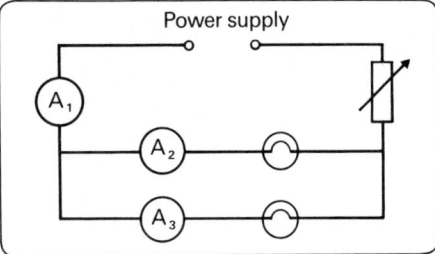

a Connect the circuit shown and adjust the rheostat to light the lamp to normal brightness. Omit ammeters A_2 and A_3 for the moment.

b Using the one ammeter, move it to positions A_2, A_3 and record the currents. How does A_1 compare with the sum of A_2 and A_3?

c Repeat this procedure after adjusting the rheostat to give a new reading on A_1.

d Copy the circuit and record your results.

ACTIVITY III

a Predict what will happen if a third lamp is added in parallel to the other two.

b Write down your prediction and test it.

ACTIVITY IV   Fuses

Fuses are safety devices designed to 'blow' or burn out when the current dangerously exceeds a certain level.

Design a circuit to measure the current rating of a sample of fuse wire.

(continues overleaf)

PLANNING & SAFETY

How will you obtain a reliable result?

Wear goggles in case of accident.

Do not connect a fuse wire directly to the terminals of any equipment, to avoid heat damage.

ACTIVITY V   Short Circuits

If a wire is connected across lamp X as shown, we say the lamp has been short circuited. *A short circuit is a low-resistance bypass.*

Power supply
12V
A
12V lamps
X
Y
Short circuit – croc lead (low resistance path)

a Set up the circuit and find out what happens with a short circuit. What three things can you see happening?

b Record your observations.

c Explain your observations, keeping in mind that the total current in the circuit is the same everywhere.

d Predict, then test, the effect of shorting out the ammeter.

ANALYSIS

Copy and complete the following.

1 A short circuit is a . . . resistance bypass.

2 In the circuit, lamp X is bypassed by a wire of very . . . resistance.

3 Lamp X goes out because it is easier for the current to flow . . . .

4 Because lamp X is shorted out, the resistance of the whole circuit is . . . (increased, reduced).

5 Lamp Y becomes brighter because, after the short, there is less . . . .

6 If a household appliance develops a short circuit, its plug fuse blows because . . . .

7 Suggest two ways in which a short circuit might occur in the flex of an electric iron.

ACTIVITY VI

a   Set up the circuit shown using two SPDT (two-way) switches, X and Y.

b   Test the circuit by switching lamp A on and off using switches X and Y.

c   Copy the diagram; explain how the circuit works. Tell the function of lamps B and C.

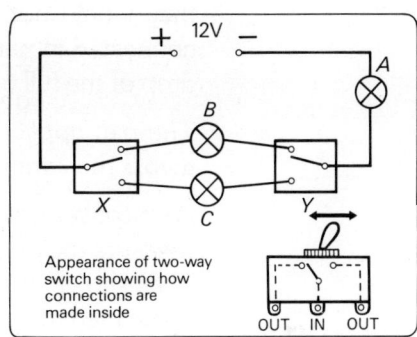

Appearance of two-way switch showing how connections are made inside

OUT   IN   OUT

4   POTENTIAL DIFFERENCE

A battery is a source of energy, stored by the chemicals inside it. As electrons flow through the battery they receive potential energy which they will use up as they pass around the circuit. *We say that there is a potential difference (p.d.) between the terminals of the battery. This potential difference is measured in volts (V), using a voltmeter.*

A 6 volt battery gives out 6 joules of energy per second when a current of 1 ampere is flowing. It has a p.d. of 6 volts between its terminals.

If a resistance wire is connected to the battery, there will be a p.d. of 6V between the ends of the wire and this will cause an electric current to flow through the wire. As the electrons flow they will lose energy and the wire will heat up. Half way along the wire, the electrons will have lost half their energy, so that the p.d. between the middle of the wire and either end is 3V. If the wire was 6 cm long there would be a p.d. drop of 1V per centimetre length. This assumes that the thickness of the wire is the same all the way along. A voltmeter is used to measure p.d. (in volts); and if connected to two points on the wire, say 2 cm apart, it would register 2V.

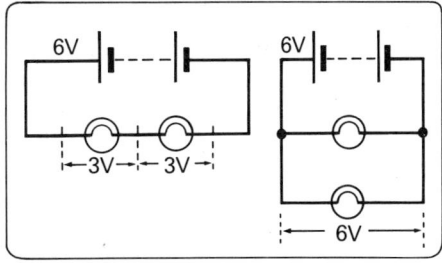

If two lamps in series were connected to the 6V battery, there would be a p.d. of 3V across each lamp. As a result if they were 6V lamps, they would not light up properly. On the other hand, if they were connected in parallel to the 6V battery, each lamp would have a p.d. drop of the full 6V across it.

The p.d. applied to a series circuit equals the sum of the p.d. drops across the various parts of the circuit.

A voltmeter is connected in parallel. To measure the p.d. between points X and Y, as shown, the voltmeter must be connected to those two points, so it must go in parallel with the resistor. To avoid causing a short circuit, the *voltmeter must have a very high resistance.*

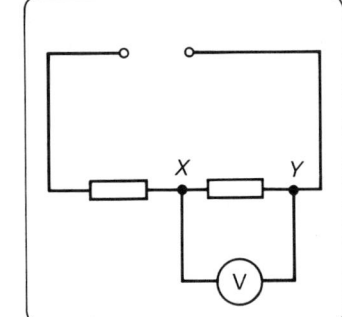

Questions

1   Copy and complete **a** and **b**.

a   A dry cell produces 1.5V. To make a 9V battery we would need . . . cells connected in . . .

b   A car battery has a p.d. of 12 volts. Therefore each of its 6 cells has a voltage of . . .

c   A car battery is faulty. One cell is not working. What would its voltage be in that case?

2   Copy this circuit and find the reading of the voltmeters. The lamps are identical.

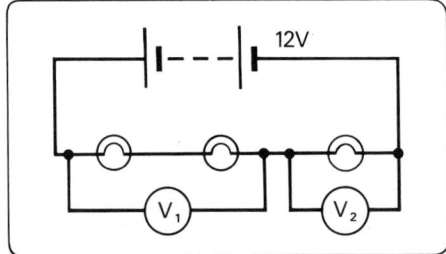

3   A 60V battery is connected to a resistance wire 20 cm long. What is the potential drop along each centimetre of wire? How could you connect the voltmeter to read 9V?

4   The diagram shows a rheostat connected as a potential divider. It is connected to a 240V supply. A lamp is connected via the sliding contact.

a   Estimate the voltage on the lamp when the slide is at A, B and C.

b   What advantage does this arrangement have over simply connecting the rheostat in series with the lamp?

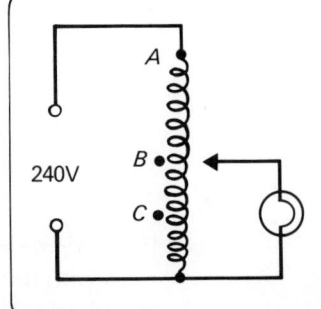

✻ 5 It takes about 20 000V to make a spark 1 cm long in air.

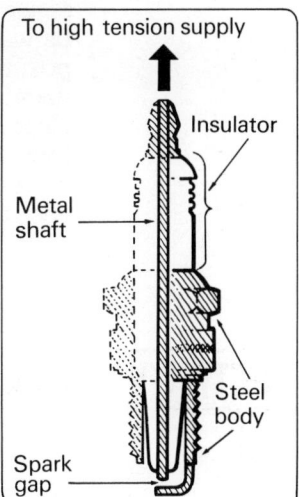

a What p.d. is needed to run a spark plug if the gap is 1 mm wide?

b Only one lead goes to the plug. How is the circuit completed?

c On a damp morning a layer of moisture may form on the insulation. How could this stop it sparking?

d A carbon deposit on the insulation near the spark gap may also stop it working. Why is this?

e Why does a plug need a higher voltage when in the engine?

6 Five identical resistors are connected to a 24V battery as shown. The voltmeter reads 3V. Find the p.d. across each of the other resistors.

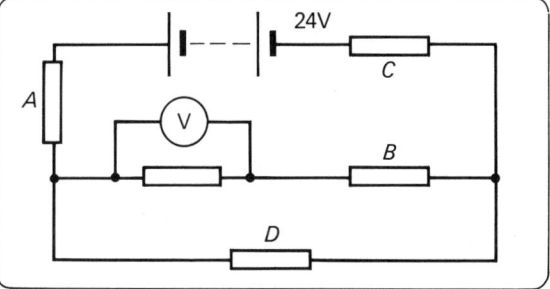

7 **Design** Suppose that a power supply is designed to give voltages from zero up to 12 volts by turning a knob and pointer on a circular dial. How would you go about checking the accuracy of the dial and how would you present your information so that it would be easily used by someone else?

8 Copy the diagram of the rheostat and use a colour pencil to mark the path taken by the current if terminals *A* and *B* are joined to a circuit. Repeat for terminals *A* and *C*. What problem will you have with the rheostat if terminals *B* and *C* are used?

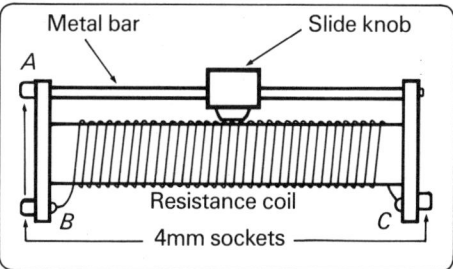

9   The diagram shows resistors connected between the positive and negative rails of an electronic circuit. Find the potential difference across each resistor.

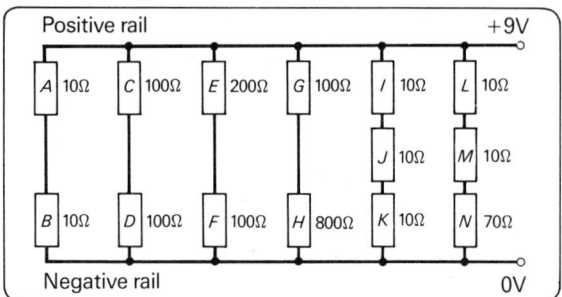

ACTIVITY I   Circuit Voltages

a   Set up the circuit shown using only the one voltmeter V_1.

b   Set the rheostat so that the lamps are very dim.

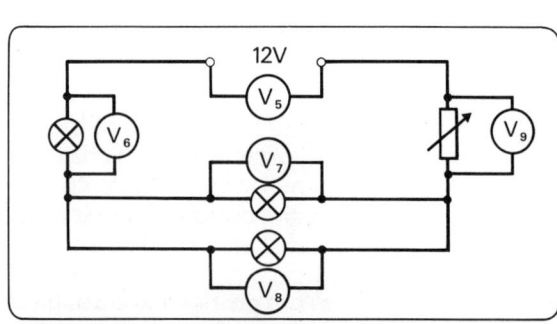

c   Measure voltage V_1;
then using the same one voltmeter, measure the p.d.'s V_2, V_3 and V_4. Record the results in a suitable table.

d   Repeat after setting the rheostat to give a new value of V_4. Or add an extra lamp to give an extra p.d. V_5.

e   What conclusions do you draw from your results?

ACTIVITY II

a   Suggest two important patterns that will be found in the voltage readings from the circuit shown.

b   Set up the circuit and collect several sets of data to test your two predictions.

ACTIVITY III Black Boxes

a Measure the voltages of various cells and batteries and 'black boxes'. Record your results in a suitable fashion.

b What conclusions can you draw about the possible contents of the black boxes?

5 CELLS AND BATTERIES

Galvani was a professor of anatomy in Italy in the 1790s. He knew that frogs' legs had some connection to electricity because he had seen them twitch when a static electricity machine was operated nearby. (Little did he realise that the machine was sending out radio waves, not discovered for another hundred years.) So, later on when the legs twitched in contact with a silver hook and a piece of iron, Galvani guessed something electrical was happening. He thought the twitching was caused by 'animal electricity' and became misled into thinking it was a new kind of electricity.

A colleague, Volta, realised that the electricity, the voltage, was generated by the two metals wet with salt solution. He soon magnified the effect by piling up pairs of disks of different metals, with blotting paper soaked in salt solution between them. His first battery or 'voltaic pile' contained about twenty disks. Later researchers like Humphrey Davy (inventor of the miner's safety lamp) had large batteries made with the metal plates a metre or more square! Although the voltage was found to depend on the number of layers in the battery, larger plates gave a longer lasting battery because they contained more chemical energy.

The ability of a cell or battery to drive current around a circuit is called the Electro-Motive Force (e.m.f.), and is measured in volts (V). A single dry cell has an e.m.f. of 1.5V, so that a 9V calculator battery contains six cells in series. A car battery has an e.m.f. of 12V, produced by six 2V cells in series. The e.m.f. of the mains supply is 240V, enough to drive a lethal current through your body.

A cell converts chemical energy into electrical energy. The millions of electrons flowing through a battery are given potential energy which they use up passing around the circuit. For example, if a lamp is in the circuit, the energy is turned into heat and light.

A battery of e.m.f. 12V gives out 12 joules of energy each second when a current of 1 ampere is flowing. We say that the battery has a potential difference (p.d.) of 12V between its terminals.

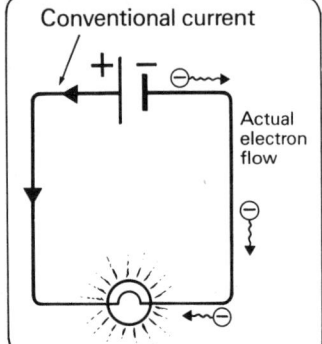

The symbol for a cell is as shown. *By convention, it was said that current flows out of the positive pole of the cell and back through the negative pole. We now know that the current consists of electrons that are negatively charged flowing in the opposite direction.*

Questions

1 a How is a battery different from a cell?

b What would you expect to find inside dry cell batteries of these voltages: 6V, 9V, 90V?

c How many car batteries could equal the voltage of the mains (240V)?

2 Suppose you have two 1.5V cells, one small and one much larger. What is the difference between them? Why are different sizes like this manufactured?

3 Investigation Build a voltaic pile using squares of two different metals separated by pieces of paper towel soaked in an electrolyte (conducting liquid) such as ammonium chloride or salt solution.

a How will you clamp the pile together?

b Test the e.m.f. of one pair of electrodes.

c Predict the number of cells needed to light a lamp such as a 1.25V type.

✳ 4 A physical data book gives the following table of Electrode Potentials, relative to hydrogen.

ELEMENT	POTENTIAL
Lithium	−3.04
Magnesium	−2.37
Aluminium	−1.71
Zinc	−0.76
Iron	−0.41
Lead	−0.13
Copper	+0.36
Silver	+0.80

a   What do you think the term electrode potential means?

b   What predictions could you make from the table?

c   How does the data from the activity agree with this accurate data?

5 a   For some applications, dry cells are connected in parallel as well as series. Why is this?

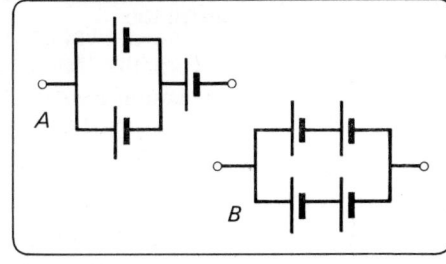

b   What is the e.m.f. of each battery shown?

c   Comment on the performance you would expect from the two batteries.

6   A solar cell generates an e.m.f. of 0.4V by converting the energy of sunlight into electrical energy.

a   How could you use solar cells to recharge a 2V lead-acid storage cell?

b   Given six solar cells, how would you use them to run a device needing 1V to operate?

ACTIVITY

The generation of an e.m.f. by pairs of metals is easily investigated with a voltmeter.

a   Set up the apparatus shown to measure the e.m.f. generated by two different metals such as zinc and copper.

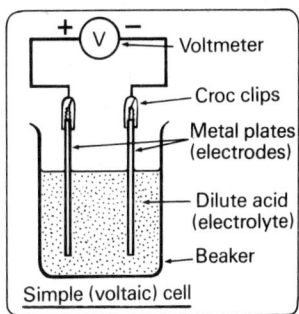

Simple (voltaic) cell

b   Find out which metal electrode is positive and which is negative. If the positive is connected to the positive of the voltmeter, then the meter reads normally. If the connections are reversed the needle goes backwards.

c   Repeat using various combinations of electrode materials including carbon (a non-metal) and magnesium.

d   Record the results in a table:

Positive electrode	Negative electrode	E.m.f. (V)

ANALYSIS

1 Which pair of electrodes gave the biggest e.m.f.?

2 Make a list of the materials tested, so that each material would be positive if used with any material below it in the list. Then label the lowest as zero potential, and add a scale showing the e.m.f. to be expected from pairing each other material with it. Use the list to predict e.m.f.s for combinations you did not actually test.

3 What energy change takes place in a cell?

6   THE LEAD-ACID ACCUMULATOR

About 1850, Sir William Grove discovered the principle of the modern lead-acid storage battery. After the invention of the voltaic pile, much work was done passing currents through liquids to see what kind of chemical changes could be produced. For example, it was possible to generate bubbles of oxygen and hydrogen from water in this way. It was found that after passing current for a few minutes, some metal electrodes were chemically changed and could act as a voltaic cell themselves. In other words, the electrodes were storing electrical energy.

A Frenchman, Planté, thoroughly investigated the effect and found that lead plates in sulphuric acid gave the best results. Researchers have yet to improve on his choice and every car still has a lead-acid battery.

The lead-acid cell or accumulator has the great advantage that it can be recharged and used over and over again. It is classified as a secondary or storage cell. Also, it has a very low internal resistance and so can deliver larger currents than a dry cell. The standard 12V car battery contains 6 cells. To recharge a battery, current is forced backwards through it, so the charger must have a higher voltage than the battery. To charge, the positive of the battery is connected to the positive of the charger, and likewise negative to negative.

The battery cells contain sulphuric acid and distilled water. In unsealed batteries the water evaporates and must be topped up every few months. If the level is allowed to fall to say half, then the effective area of the lead plates is likewise reduced and so is the storage capacity. As a cell charges, one plate develops a coating of lead oxide, so that the cell electrodes that generate the e.m.f. are lead and lead oxide rather than two different metals. Another storage battery is the NiFe. Perhaps you can guess what its two electrodes are made of.

When fully charged, a battery starts to 'gas', giving off bubbles of oxygen and hydrogen which form an explosive mixture. Lighted cigarettes should never be brought near batteries. Likewise sparks should be avoided by switching the charger off at the mains before disconnecting the terminals. It has been known for exploding batteries to blow the roof off a workshop.

Questions

1   What is the advantage of an accumulator?

2   What are its disadvantages (especially for electric cars that need many of them)?

3   Which do you think stores the most energy per cubic inch of space, petrol or a lead-acid battery?

4   What energy change takes place in a secondary cell, when
   a   discharging?
   b   charging?
   c   What safety precautions must be observed when checking on a charging battery?

5   **Design**   How could you design a lead-acid battery to
   a   hold more energy without becoming very bulky?
   b   have a higher voltage?

6   Capacity of car batteries is given in ampere-hours, so that in theory a 40 amp-hour battery could supply a current of 1A for 40 hours. How long could it supply current of
   a   4A?
   b   10A?
   c   How long could it supply two headlamps drawing 6A each?
   d   How many amp-hours of energy would be used up by running an 80A starter motor for 1 minute?
   e   If the car dynamo recharges the battery at 15A, how long will it take to replace this energy?

7 Battery chargers as shown use a transformer to reduce the mains voltage. The mains voltage is AC (Alternating Current reversing 50 times each second) but the battery charger needs DC (Direct Current, same direction all the time). For this reason the AC current must be changed to DC.

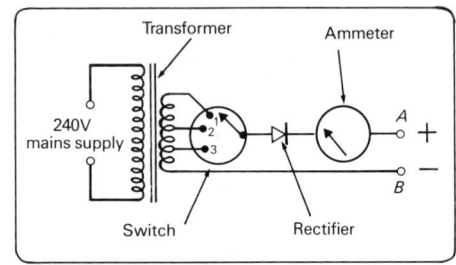

a What part of the charger do you think changes AC to DC? Use a reference book to help.

b Why are there three poles on the switch?

c A cheap 'trickle' charger is designed to charge a car battery slowly over a long time. What parts do you think are omitted for economy in making a trickle charger?

d What voltage might the transformer give out on setting 3, the lowest one?

e How would a battery be connected?

f What information about a rectifier can be inferred from the diagram?

g Predict the danger of connecting battery wrongly.

8 **Investigation**   What variables might control the amount of energy a lead-acid cell can hold? Design a controlled experiment to test one of them.

9 A car with a 55A-hr battery is parked at 8.00 a.m. and the lights are accidentally left on. Each headlamp takes 5A and each rear light about 0.5A. When the owner returns after work the clock is seen to have stopped. Suggest what time it might have stopped at.

ACTIVITY

The action of a lead-acid cell is easily investigated as follows.

Step 1

a Set up the apparatus as shown using two sheets of lead separated by a piece of paper towel and held together by rubber bands. The electrodes are placed in a beaker two-thirds full of dilute sulphuric acid.

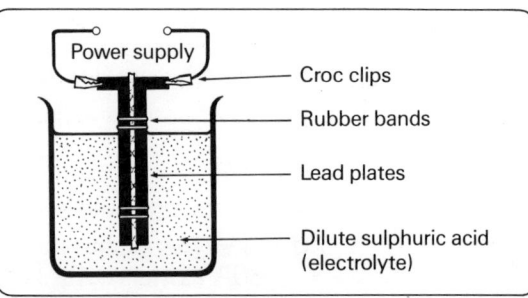

b Connect the cell to a power supply set on 3 or 4V. Mark the electrode connected to the positive. Charge the cell for 60 seconds.

c Disconnect the cell and check its e.m.f. with a voltmeter. Verify that the positive plate is the one connected to the positive supply.

d Try lighting a 1.25V lamp from the cell.

Step 2

a Try charging the cell for various lengths of time, increasing by 0.5 minutes each time. In each case measure the time for which it can keep the lamp glowing on discharge.

b Record the results in a table/graph. Plot a graph of discharging time against charging time.

c What conclusions can you draw from your charge/discharge graph?

---

7   OHM'S LAW

The way water flows through pipes is a good model for understanding how electricity flows through a circuit. In this model, water pressure represents voltage (potential difference). Increasing the pressure speeds up the water flow. The amount of water flowing through a pipe per second represents the current. Increasing the voltage increases the current in a circuit.

Water flows much more slowly through a long hose-pipe than a short one, because the pipe resists or slows down the flow. In the same way a long wire has more resistance to current than a short one; and a thin wire (like a narrow pipe) has more resistance than a thick one.

The idea of electrical resistance was put forward in 1827 by George Simon Ohm. He used thermocouples to drive current through wire samples. A thermocouple can be made from two wires of different metal, twisted together at one end. When the junction is heated, a voltage is generated between the other ends. By using two or three thermocouples in series he could double or triple the p.d. applied to a wire. He found that the current produced (I) was doubled and tripled, so that it was directly proportional to the p.d. (V) for a given wire at a fixed temperature. The ratio of voltage to current was called the resistance (R). *Ohm's law can be written using symbols as:*

$$V = IR, \qquad I = \frac{V}{R}, \qquad R = \frac{V}{I}.$$

If V is in volts and I in amperes, then resistance, R, is in ohms (Symbol Ω, Greek letter Omega or 'O').

The resistance of metals increases with temperature, but semiconductors such as carbon have less resistance when hot. Metal alloys have been developed that show very little change in resistance with temperature.

Ohm's law may be stated as follows: The current in a conductor is directly proportional to the potential difference applied to it, if the temperature is fixed.

Example: A lamp draws a current of 3A from a 12V battery. Find its resistance.

Working: By Ohm's law: $R = \frac{V}{I}$. Therefore $R = \frac{12}{3} = 4$ ohms.

In electronic circuits very high resistances are used, measured in kilohms $(k\Omega)$ and megohms $(M\Omega)$.
$1\,k\Omega = 1000\Omega$ and $1\,M\Omega = 1\,000\,000\Omega$.

The statement of Ohm's law says that the temperature must be constant. If the temperature rises, the resistance of a metal will increase. This is why the resistance of the filament of a lamp is much higher when incandescent. On the other hand, semiconducting materials such as silicon and carbon have a negative temperature coefficient. Their resistance reduces greatly as they warm up. Such resistors are called thermistors.

In 1911 a scientist named Kamerlingh Onnes made an important discovery when investigating the effect of temperature on resistance. He found that mercury suddenly lost all its resistance at about $-253°C$ (20 Kelvin) and became a 'superconductor'. Other materials have now been discovered that superconduct at much higher temperatures than this. If researchers can develop materials that superconduct at normal temperatures, tremendous improvements in our technology will become possible. One basic application would be superconducting power cables. Another could be extremely powerful electromagnets which would keep their strength without a power supply.

Questions

1 Use Ohm's law to calculate the values missing from the table.

DEVICE		VOLTAGE (V)	CURRENT (A)	RESISTANCE (Ω)
a	Car headlamp	12	5	–
b	Hair drier	240	4	–
c	Fan heater	240	–	20
d	Flashlamp	–	0.06	100
e	Car starter motor	12	–	0.125
f	Nightlight	240	0.06	–
g	American electric kettle	–	10	11.5
h	Resistor in TV set	–	0.45mA	20k

2 For the circuit shown:

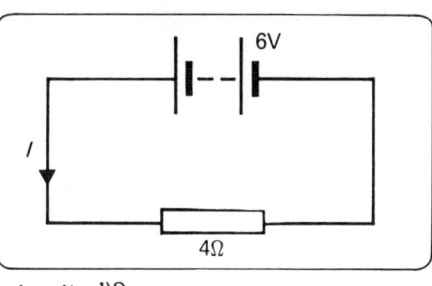

a What current will flow?

b What if the battery itself had a resistance of 2 ohms?

c What current would flow if the resistor was replaced by a thick copper wire (if the battery was short-circuited)?

3 For the circuit shown:

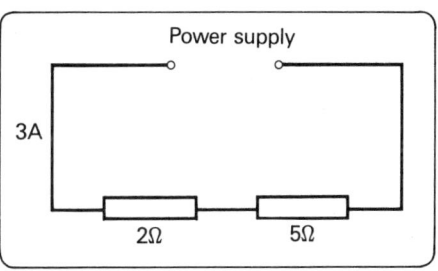

a What is the total resistance of the circuit?

b What is the e.m.f. (voltage) of the power supply?

c Find the p.d. across each resistor.

d What resistance must be added to reduce the current to 0.1A?

4 **a** A lamp takes 0.25A from the 240V supply. Find the resistance of the lamp.

b A car headlamp takes a current of 3A. Find its resistance.

c A car starter motor draws a current of about 80A. What is its resistance?

d A 120Ω resistor is connected to the mains. What current will flow?

5   Design You want to make a 1 kilowatt heater, using wire of resistance 5 ohms/metre. It must take a current of 4A from the 240V supply. Find its resistance. Explain how you could make it.

6   a A mains lamp has a resistance of 960 ohms. A boy checked it using a 6 volt battery plus an ammeter. What did he find?

b An engineer found that when a 60W lamp was switched on there was a surge of current up to 3A before it fell back to the normal 0.25A. What was he able to work out from this?

7 A car headlamp bulb takes a current of 2A.

a Find its resistance.

b Suppose you connected it to the mains. What resistance would need to be in series with it to limit its current to 2A?

8 When the circuit shown was switched on, the ammeter needle swung to 2.1A, then quickly fell to 2.0A and stayed there.

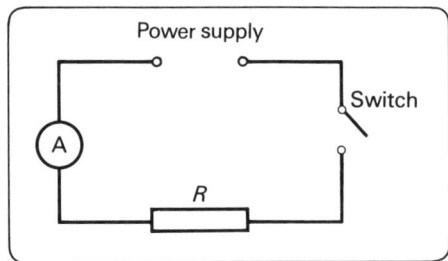

a What could be the reason for this?

b What would happen if the resistor was put in a flame?

c With a different resistor made of an alloy called constantan, the ammeter needle went straight to 2.0A and stayed there. What can you infer from this about constantan?

d Using a third kind of resistor the needle swung to 2.0A then gradually crept higher. What does this tell you about the material this resistor was made of?

9 Display the following information on resistivity in graphical form. Comment on it.

MATERIAL	RESISTIVITY (UNITS)
Aluminium	2.7
Carbon	1000
Copper	1.7
Gold	2.3
Iron	10.5
Lead	21.0
Silver	1.6
Mercury	96.0

10 **Investigation**   The resistance of a thermistor decreases as it gets hot.

   a How could you measure its resistance at various temperatures?

   b How could you use a thermistor to measure temperature?

11 **Design**   An ohm-meter is used to measure resistance. It contains a milliammeter and a battery, and has two leads to connect to the resistor being tested. How could you make and calibrate an ohm-meter?

✳ 12   The diagram shows part of an electronics circuit. It is a potential divider.

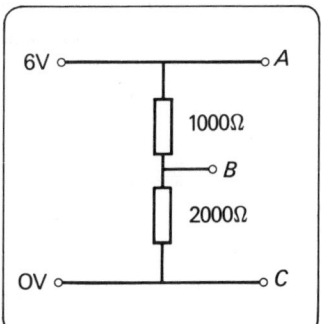

   a Find the total resistance between A and C.

   b What current will flow? Give your answer in mA.

   c Find the p.d. between A and B, and between B and C.

✳ 13   If two resistors are connected in parallel, the resistance of the circuit is reduced, because having two resistors makes it easier for the current to flow. Similarly, adding a second road between two towns makes it easier for the traffic to flow.

The combined effect or equivalent resistance, R, can be worked out using this formula: $\dfrac{1}{R} = \dfrac{1}{R_1} + \dfrac{1}{R_2}$. Work out the combined resistance of the following in parallel.

   a Two 10 ohms

   b Two 1000 ohms

   c One 10 ohms and one 1 ohm

   d One 100 ohms and one 10 ohms

   e One 3 ohms and one 5 ohms

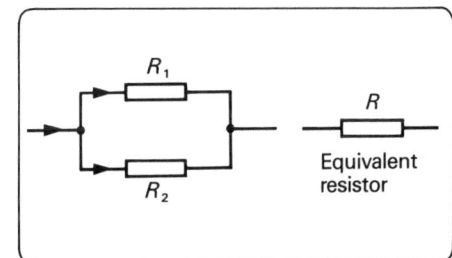

   f What conclusion about the combined resistance can be drawn from your calculations?

✳ 14   **Design** **a** You are given a 0–5V voltmeter and a 10000 ohm (10kΩ) variable resistor. How could you adapt the voltmeter to read 0–10V?

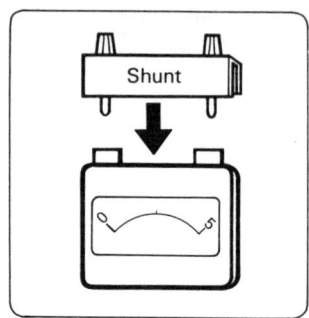

b   The range of a voltmeter can be changed by fitting a 'shunt' onto its terminals as shown. Explain what would be found inside the shunt.

ACTIVITY I   Resistance of a Wire

Ohm's law can be verified with a simple circuit as shown.

a   Set up the circuit shown where resistor R is a 50 cm length of constantan wire.

b   Connect the resistance wire between crocodile clips and carefully measure the length between the clips and record it.

c   Set the rheostat to give the smallest current you can measure. Record the current (I), and p.d. across the wire (V).

d   By adjusting the rheostat, obtain at least four more pairs of values for current and voltage. Obtain the widest range possible with your apparatus. Switch off current between readings.

e   Record your results in a table.

V (volts)	I (amps)	R (ohms)

f   *Graph*   Plot a graph of voltage against current. Draw the best line through the points. It should be a straight line.

ACTIVITY II   Resistance of a Lamp Filament

Repeat the experiment using a 12V lamp in place of the resistance wire. The graph will be different this time.

ANALYSIS

1 What conclusions can you draw from the graphs for the two experiments?

2 Which points on your graphs are the most reliable, those at the lowest currents or those at the higher currents? Explain why.

3 What was the resistance per metre of the wire?

4 What length of this wire would you need to make a 50 ohm resistor?

5 What special property does constantan alloy have?

6 A similar experiment gave the graph shown. What can you infer from it?

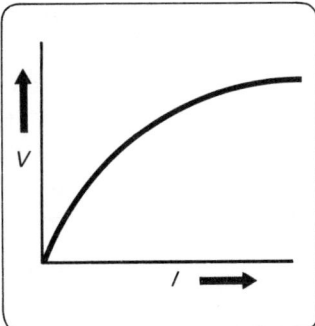

8   ELECTRIC POWER

Electric currents are used to carry energy from one place to another. For example, on a bicycle, energy is carried from the dynamo to the lamp by electric current flowing through wires (along one wire and back through the metal frame). As the electrons pass through the dynamo they are given potential energy. Some of this energy is used up as they pass through the wires, but most of it is turned to heat and light in the lamp filament.

In a battery, chemical energy is used to give potential energy to the electrons. A 12V battery gives out 12 joules of energy each second when a current of 1A is flowing. This energy is carried away by the electrons. A current of, say, 3A would carry away 3×12 or 36 joules of energy each second. Since power is the amount of energy used per second, the power output of the battery would be 36 watts, 36W. The conclusion of this is: *Electric Power (P) is calculated by multiplying current (I) by voltage (V). So that we have the formula $P = IV$ watts, which may also be written $P = I^2 R$ and $P = \dfrac{V^2}{R}$.*

Note that if a battery is supplying energy at the rate of 36W, it is also being consumed in the circuit at this rate.

Example: A spotlamp draws a current of 6A from a 12V car battery. Find the power of the lamp.

Working: $P = IV$. Therefore $P = 6 \times 12 = 72$ watts.

Example: An immersion heater to heat a single cup of water for making coffee has a resistance of 2 ohms and plugs into the cigarette lighter of a car. Find its power.

Working: $P = \dfrac{V^2}{R}$. Therefore $P = \dfrac{144}{2} = 72$ watts.

A 60W lamp uses 60 joules of energy each second. A 500W electric motor does 500 joules of work each second. Fan heaters are rated in kilowatts (kW), commonly 2 or 3kW. A 3kW heater supplies 3000J of heat energy each second and draws 3000J from the mains each second.

Car headlamp bulbs are marked with information such as 75W – 12V. This means that the lamp uses 75J of energy each second if run off a 12V supply. If connected to the 240V mains, the power would be so large that the lamp filament would instantly melt. Notice that ordinary incandescent lamps are very inefficient, and well over 90% of the energy used turns to heat rather than light energy. Fluorescent strip lamps are far more efficient and have much lower wattage rating for the same light output.

Before investigating the power of a lamp, work through these power problems.

Problems

1   A car headlamp takes 3A from a 12V supply. Find its power and resistance.

2   A house lamp takes 0.25A from a 240V supply. Find its wattage (power) and resistance.

3   Find the current taken from the mains by a 720W hair drier.

4   Find the current taken from the mains by a 3kW heater. What would be the minimum value fuse in the plug of such an appliance (2A, 5A, 10A, 13A)?

5   A 15W night lamp runs off the 240V mains. Find its current and resistance.

6   A 6 ohm resistor is connected to a 12V battery. Find the power output of the battery. What happens to this power?

7   A car starter draws a current of about 80A. What is the power of the motor? If there are 746W in the old 1 horse-power unit, what is the HP of the motor?

8   An engineer is designing a 2kW heater using wire of resistance 5 ohms/metre. It is to work off the 240V mains. What length of wire will he need?

9   A car has on two 75W headlamps, two 15W rear lamps and the driver brakes to light up two 30W brake lights. What current will flow from the battery? Why are brake lights more powerful than rear lamps?

10   **Design**   In an electric car, energy is stored in a battery rather than in petrol. This question compares the two storage systems. A typical 12V car battery has a storage capacity of 40 ampere-hours and in theory can deliver a current of, say, 5A for 8 hours.

    a   What is the power output of a battery in watts for a current of 5A?

    b   How many joules of energy will it give out in one minute?

    c   Calculate the energy given out during the 8 hours discharge.

    d   Suppose the capacity of a car petrol tank is 50 litres and petrol has a density of 0.8 kg/litre. What mass of petrol is carried?

    e   If the calorific value of petrol is 20mJ/kg, how much energy is stored in the tank of petrol?

    f   How many batteries would be needed to store the same amount of energy? Estimate their total mass.

    g   How would this problem be minimised if the electric car was just for local commuting?

11   A 72W car headlamp and a 240W mains lamp are connected in series to the mains. What will happen?

✷ **12**   A solar cell converts energy from the sun into electrical energy.

    a   If a solar cell measuring 10 cm by 10 cm generates an e.m.f. of 0.4V and delivers a current of 2A, find its power output.

    b   How many cells would be needed for a solar panel of area 1m²?

    c   Find the power output of such a panel.

    d   If solar energy reaches the Earth's surface on a hot day at the rate of 1kW per square metre, find the efficiency of the panel, the percentage of radiant energy it converts to electrical energy.

ACTIVITY

    a   Set up the circuit shown so that the current and voltage for a car lamp bulb can be measured over a wide range of values.

    b   Start with the lowest value of current and voltage you can measure.

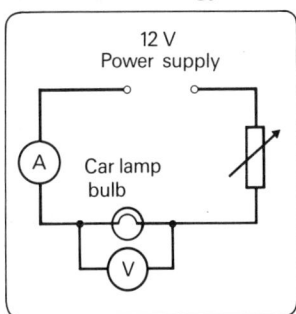

12 V
Power supply

A   Car lamp bulb

V

c   Increase the voltage and current by steps until you reach the normal operating voltage of 12V. Record values of current and voltage in a table as shown. Compute the power consumption for each voltage.

V (volts)	I (amps)	P = IV (watts)

d   *Graph*   Plot a graph of power (watts) against current (amperes). Draw a smooth curve through the points.

ANALYSIS

1   From your graph, what was the power of the lamp at 6V?

2   What was the power at 12V? How does this compare with the stated wattage of the lamp? Express the difference as a percentage of the stated value.

3   From the graph, predict what the power would be at 24V.

4   Using the formula $P = \dfrac{V^2}{R}$, how should the power at 24V compare with that at 12V?

5   Why would the lamp not work on 24V?

Questions

1   **Library**   a   Find and make a copy of a diagram of an incandescent lamp. Explain how it works.

b   The filament is made of tungsten (chemical symbol **W**). Find out why this metal is so suitable for the job.

c   What property would the ideal lamp filament possess?

2   **Library**   Thomas Edison was a genius, but when people asked his secret he said that genius was '1% inspiration and 99% perspiration'. What did he mean by this? Find out how his work on the first light bulb illustrated the truth of this statement.

3 **Investigation** Electric motors have the important property that they respond to heavy loads by pulling harder instead of getting weaker like petrol engines. As a result they draw more current (which is why windscreen wipers frozen to a windscreen will quickly overload and burn out if not set free).

    a Design an experiment to investigate how the power consumption of a small electric motor depends on the load it has to lift on its flywheel.

    b What measurements will you need to monitor the electrical power input to the motor?

    c To make it a controlled experiment, what should you do to the voltage of the supply?

    d What precautions will you take to avoid burning out the motor with loads too heavy for it to lift?

    e How will you best display your results?

    f How could you extend the experiment to measure the mechanical power output (the work done each second lifting the weight)?

4 **Assignment** Find the power ratings of at least five electrical appliances in your home.

✱ 5 **a** The Lucas 15AR alternator on vehicle X can generate 40A of current at a voltage of 14.5V. Calculate its power output.

    b The starter motor on vehicle Y draws a current of 100A. Calculate the power consumption of the motor.

    c Vehicle Y has a flat battery and so jumper cables are used to connect its electrical system to that of vehicle X. How is it possible for this arrangement to work?

    d What additional information may be calculated from the data given?

6 **Design** An engineer is studying the feasibility of making an electrically heated overall or suit for old people to wear in the home in winter. It would run off a 40 amp-hour car battery.

    a Suggest a suitable power value for the heating element.

    b How many hours a day could it run off the battery?

    c Calculate the resistance of the element.

    d How long would the resistance wire (element) have to be to spread heat over arms and legs as well as body?

    e In searching in a catalogue for suitable wire, what specification would you have in mind?

9 COST OF ELECTRICITY

After running a fan heater for an hour there is no more 'electricity' in your home than there was before, so what are you paying for when the electricity bill comes? You are paying for the energy that was carried to your home by electric current. It is the energy itself you pay for, just as you pay the dairy for the milk, not the bottles it came in.

Payment for electrical energy is based on the old Board of Trade unit, also called the kilowatt-hour or simply a 'unit'. A unit of electricity is the energy used by a 1 kilowatt appliance in 1 hour. So that a 3kW immerser running for 2 hours will consume 3×2, or 6 units of energy. At present the cost per unit is about 5 pence, so that the cost of running the 3kW heater for 2 hours would be 6×5, or 30p.

A 100W lamp uses $\dfrac{1}{10}$ of a unit per hour, so that in 20 hours it would use $20 \times \dfrac{1}{10}$ or 2 units. The cost would be 2×5 or 10 pence.

The electricity meter records the number of kilowatt-hours (units) used. Usually the meter is read and a bill sent every three months, about 90 days. So that the running of the 3kW immersion heater mentioned above for two hours a day at a cost of 30p would add £27 to your electricity bill. The Economy 7 system provides half-price electricity for seven hours during the night and is mainly used for off-peak heating of hot water and storage heaters. Economy 7 is an attempt by the electricity board to spread some of the generating load from day to night to reduce the need for extra equipment. This is necessary because people are using more and more appliances and somehow the power must be available when they plug in.

Questions

1   The chart shows the power ratings of some common appliances. Copy and complete the table to find the cost per hour for operating each one, at the cost per unit given by your teacher.

| APPLIANCE | POWER | | COST /HOUR |
	Watts	kW	
Fan heater	3000		
Kettle	2200		
Hair drier	800		
Lamp	60		
Vacuum cleaner	500		
Iron	750		
Tumble drier	2700		
Shower	8000		
Toaster	1100		
Night light	20		
Microwave oven	660		
Paint stripper	1600		
Jig saw	350		

2 An electricity bill had the following information:

Meter readings: 84119, 84900E.

Cost per unit: 5p

Quarterly charge: £7·00

a How many units were used?

b Find the total cost including the quarterly charge.

c What is the purpose of the quarterly fixed charge?

d The meter is often inside the house and is read by a person who visits every house each quarter. Why is there an 'E' by the second meter reading?

3 An 8kW electric shower is run for 6 minutes.

a Find the number of units used.

b Find the cost at 5p per unit.

4 **a** Find the cost of running a 2kW fan heater for five hours at 5p a unit every day during the billing period.

b Estimate the cost of leaving a 100W outside light on all night for security purposes.

5 A 3kW immersion heater runs for 5 hours each night on the Economy 7 tariff at a cost of 2.5p per unit. Find the cost for the quarter.

✳ 6 A TV set has a power of 80W.

    a Find the cost of viewing per hour.

    b Estimate the cost of one week's viewing.

    c How does this compare with the cost per week of the TV licence?

    d If a TV cost £300 and lasted 10 years, work out its weekly contribution to the cost of viewing.

7 Estimate the cost of boiling water in a 3kW electric kettle to make tea.

8 A battery charger draws a current of 0.1A from a 240V supply. If it takes ten hours to charge a car battery, find the cost at 5p/unit.

9 A woman uses a 25W heater belt to keep a gallon of wine warm, to speed up the fermentation.

    a Find the operating cost per hour.

    b If the process takes 3 weeks, estimate the cost of the electrical energy at 5p/unit.

    c Find the annual cost if she ferments a gallon each month.

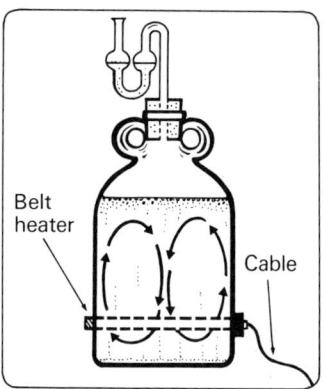

    d A thermostatically controlled heater costs about £5 and keeps the wine at the ideal temperature, although it spends half the time turned off. Would it be a good investment for the woman?

    e Why does a thermostatically controlled immersion heater need less energy to keep the wine at the same temperature as the belt heater?

    f Copy the diagram then add a second one showing a rod-shaped immersion heater fitted at the top.

10 The diagrams show two electricity dial meter readings, the start and end of the quarter. Read the meters and work out the bill at 5p/unit.

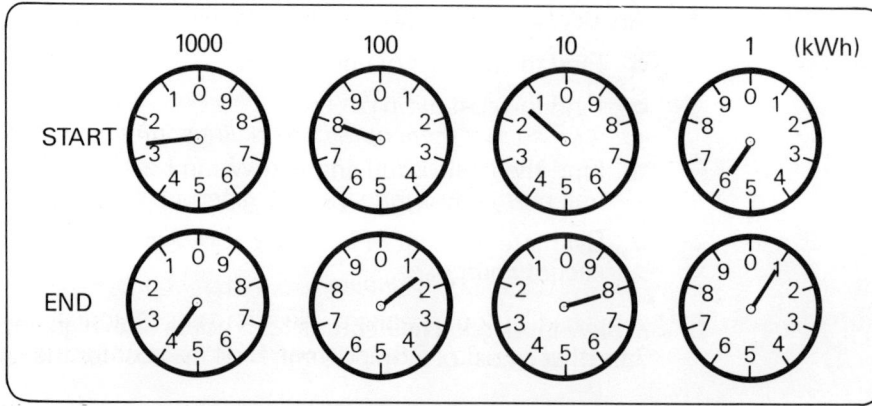

11   The electricity board main fuse in a house is typically a 60A one. Imagine that a fault occurred in a house while the owner was away, and that current was drained at the maximum rate, day and night. Estimate the cost per hour, day, and billing period at 5p per unit.

10   MAGNETISM

Weak, natural magnets were used by the ancient Greeks, two thousand years ago. These magnets were lumps of iron ore and were named magnets after the district of Magnesia (in modern Turkey) where the ore deposits were. Greek shepherds found that their iron crooks were weakly attracted to certain rocks.

The lumps of ore, magnetite, were later called lodestones (direction-stones). The fact that a suspended or pivoted magnet would point north and could be used as a compass was not discovered until the sixth century. By AD 1100 the Chinese were using compasses to navigate.

About AD 1260 Peter the Pilgrim, a French crusader, wrote a book about magnets. He knew they had two poles, that breaking a magnet in two created two magnets each with two poles, and that a magnet could turn a piece of iron nearby into a magnet (induced magnetism).

By 1600, William Gilbert had realised that the Earth was also a magnet and that a compass worked because it was attracted by the two magnetic poles of the Earth. In fact he made a model Earth, his 'terrella', of lodestone to illustrate his ideas. There were many superstitious ideas about magnetism, but Gilbert was a methodical scientist. All he wrote was based on careful experiment: things which he had 'explored and many times performed and repeated'.

Fine particles of iron filings, scattered around a magnet will form a pattern, so that there appear to be lines running from one pole to the other. Michael Faraday called these 'lines of force'. As we shall see, the lines of force model is extremely useful in explaining magnetic effects.

Today we are able to make strong magnets using electricity. Magnets are used in loudspeakers, electric motors and dynamos. Magnetic tapes (magnetic materials coated on plastic tape) are used to record sound in cassettes, and the floppy disks for computers also store information magnetically. Both cassette tape and floppy disks will be damaged if brought near magnets.

Only a few materials are strongly magnetic. They are the metals iron, steel (an iron alloy), cobalt, and nickel. Ferric oxide (made from iron) and chromium oxide are also magnetic. These are used on cassette tapes. *Bar magnets are made of various iron alloys, and researchers are constantly testing new ones. The names of the alloys, such as Alnico, give a clue to their content in addition to the iron. Ferrites are also iron compounds and are moulded to make magnets of various shapes.* A magnet can be used to test for magnetic materials.

ACTIVITY I

a Test as many materials as possible with a magnet to see which are attracted.

b Record your results, listing by material rather than name of object tested.

ACTIVITY II

Use paper clips to find out where a magnet attracts most and least.

Questions

1   What conclusions do you draw from your testing?

2   Would it be true to say that most metals are magnetic?

3   Would it be true to say that most metal objects are magnetic?

4   Tin is not magnetic. Why is a 'tin' can attracted by a magnet? Explain the difference between the true 'tin can' and 'tinned can'.

5   What did William Gilbert discover?

6   **Investigation**   Could paper be very weakly magnetic? How could you make a more sensitive test than simply trying to pick a piece up with a magnet?

7   **Investigation**   A magnet does not seem to attract a piece of iron several centimetres away. Is this because the force gets weaker with distance or because the air gradually screens off or absorbs the magnetic force? How could this be investigated?

11   MAGNETIC FIELDS

A magnet attracts iron objects nearby with a force that gets weaker with distance. To explain this 'action at a distance' *Michael Faraday imagined the space around the magnet to be filled with 'lines of force', running from its north pole to its south pole. We say there is a magnetic field around a magnet and that where it is strongest the lines of force are closest together.* According to Faraday's model, the lines of force repel one another and never cross or touch. The experiment that gave Faraday the idea is easy to repeat.

ACTIVITY

Step 1   Identifying the north (N) and south (S) poles

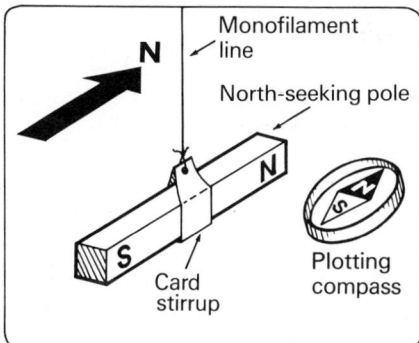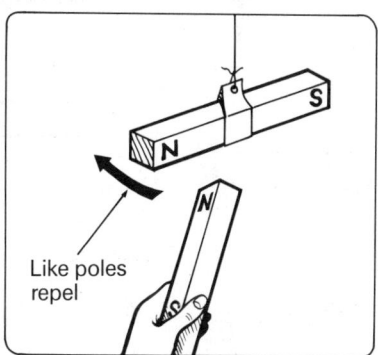

Suspend a magnet on a stirrup (perhaps made of card) as shown so that it is free to rotate. The end that finishes pointing north is technically called the 'north-seeking pole'. Mark it in some way to identify it.

Identify the poles of another magnet by using the 'law of poles' (that two N poles repel, while a N and S attract), so that a pole attracting your known N pole must be a south.

Step 2   Field of a bar magnet

Cover a bar magnet with a sheet of paper.

Gently sprinkle iron filings over the paper. Tap the paper lightly to help the filings line up.

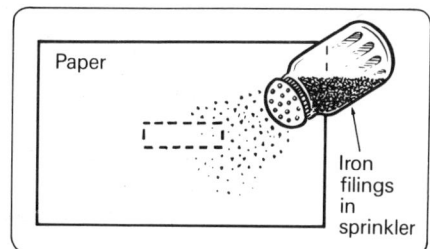

Which of the field patterns below best represents the one you see?

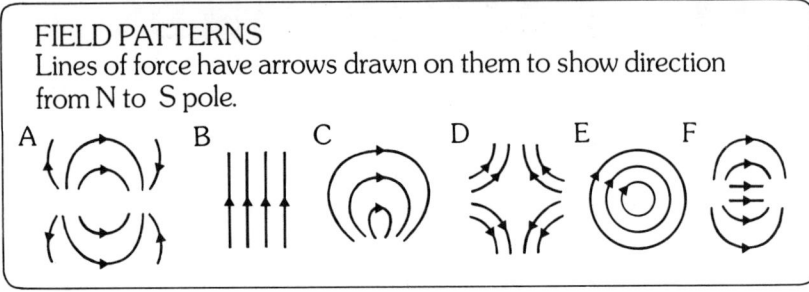

FIELD PATTERNS
Lines of force have arrows drawn on them to show direction
from N to S pole.

Step 3
Test each of the pole arrangements below. In each case identify the
pattern from that above. (One is missed out.)

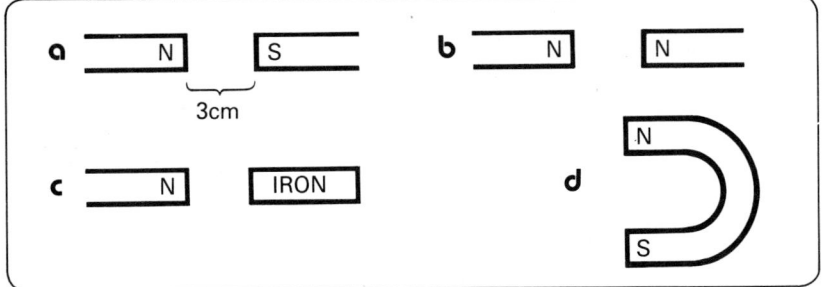

Step 4
Investigate the pole positions and fields of several squares of steel
sheet that someone has rubbed with a magnet.

Questions   Copy the pole arrangements you tested and draw in the magnetic field
patterns.

1   List the properties of the lines of force in Faraday's model.

2   What does the law of poles state?

3   How could you obtain the magnetic
pattern shown?

4   A small compass (plotting compass)
points along the direction of the
magnetic field. Predict the direction the
compass will point in at positions A, B
and C as shown. Use an arrow in a circle
to represent the compass with the arrow
head as the north-seeking pole. (Hint:
Draw in lines of force.)

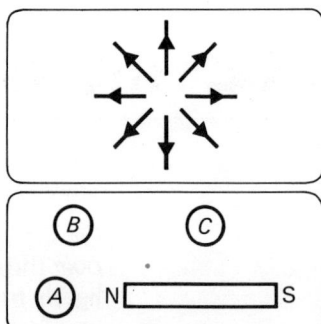

5 A magnet is moved sideways towards a plotting compass, as shown.

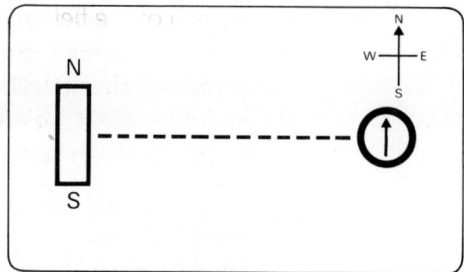

 a Predict what will happen and why.

 b How could this procedure be used to compare strengths of magnets?

 c What would happen if the magnet was reversed?

 d Find out from a reference book what a 'neutral point' is.

6 **Library** **a** Use a resource book to find a diagram representing the Earth's magnetic field. Make a copy.

 b What is 'angle of variation' and why is it important in navigation?

 c What is 'angle of dip'?

✱ 7 If you were fitting a car compass, what problems might you expect? How would you test to be sure it was working properly?

8 The strength of a magnetic pole may be tested in several ways, such as picking up paper clips or iron filings, and by deflecting a compass needle.

 a In which tests is the force of gravity involved?

 b What force is involved in the other test?

 c Which test is least sensitive and why?

 d If you had a magnet and an iron bar, how could you tell them apart without using any kind of apparatus?

 e If the north pole of a magnet pulls towards a metal bar, this does not prove the bar is a magnet. Explain why. What would be the sure proof that the bar was a magnet?

12 ATOMIC THEORY OF MAGNETISM

According to the atomic or molecular theory, each atom of a magnetic material is already a tiny magnet, having a north and south pole. In a magnet, these atomic magnets are aligned, producing north poles at one end of the bar and south at the other. In an unmagnetised bar the atomic magnets are pointing in all directions.

This is illustrated in the diagrams, where small arrows represent the atomic magnets, with the arrow heads as north-seeking poles.

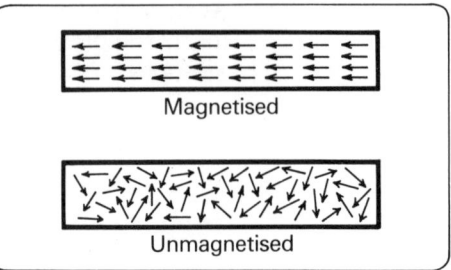

One way to magnetise a steel bar is to stroke it in one direction with a strong magnet. This causes the atomic magnets to align. When the atomic magnets are fully aligned the magnet reaches saturation and can be made no stronger. Striking a magnet and strong heating help to destroy the alignment and so weaken the magnet. *Although steel can form permanent magnets, pure iron cannot. It is said to be magnetically soft and only stays magnetised while in a magnetic field.* The field aligns the atomic magnets by a process called induction. This is shown in the diagram. By drawing in the atomic magnets it is easy to show that poles shown induced in the iron are correct.

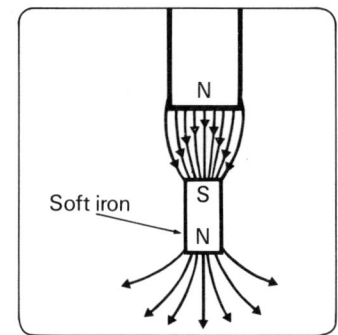

Nobody has ever made a magnet with only one pole. If a magnet is cut or broken as shown, new magnets will be formed, each with two poles. By drawing in atomic magnets, it is easy to verify that the poles shown in the diagram are correct.

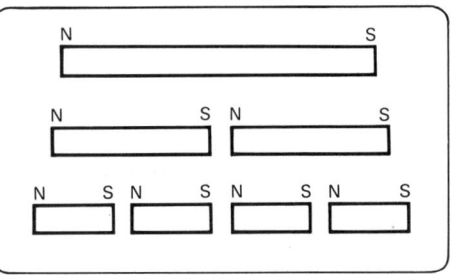

ACTIVITY

Step 1

Magnetise a strip of hacksaw blade by stroking with a magnet as shown. Predict what pole will form at A and test it with a plotting compass.

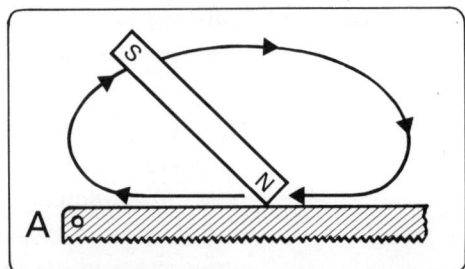

Step 2
Try magnetising a bar of soft iron by stroking. Can you magnetise it strongly enough to pick up a paper clip or iron filings?

Step 3
The diagram shows how an iron bar may be magnetised by induction. Test to see that the poles shown are correct.

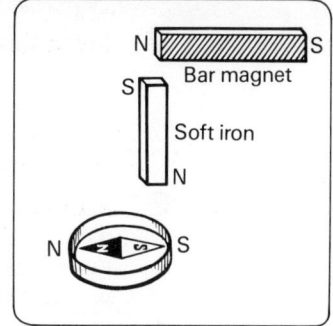

Step 4
Magnetise a strip of hacksaw blade by stroking with a magnet. Snap the blade and test the new poles formed. Are they as predicted by the atomic theory?

State your conclusions for each activity.

Questions

1 Copy each diagram above and draw in atomic magnets where necessary. Add notes explaining what each shows.

2 Explain what magnetic induction is.

3 Would it be possible to magnetise a steel nail strongly enough to lift a car or tank? Explain.

4 In electromagnets, why is the wiring wrapped around an iron core rather than a steel one?

5 When a crucible of molten steel was cooled down, the steel was found to be weakly magnetised. Explain why and predict the location of the poles.

6 **Investigation**   The diagram shows a compass needle near a small magnet.

a What do arrows A and B represent?

b Why are the arrows different lengths?

c How would the diagram change if the arrows were to be of equal length?

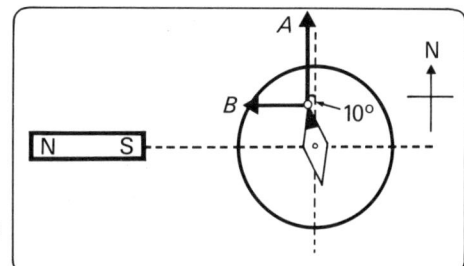

d How would the magnet be moved to cause this?

e Design an investigation to compare the pole strengths of several magnets, given a half-metre, compass and protractor.

7 **Investigation**   Is a length of steel hacksaw blade magnetised more strongly by stroking it more times with a magnet? How would you investigate this?
(A sample may be demagnetised by passing it through a coil carrying AC current.)

8 **Investigation**   If a magnetised strip of hacksaw blade was heated in a bunsen flame, would this affect its magnetism? How would you investigate this? Tell what you would do and what measurements you would make.

9 **Library**   The atomic theory described above is a simplification. Use a resource book to find out about magnetic domains.

13   ELECTROMAGNETS

Many people looked for relationships between magnetism and electricity. A clue to the connection was provided by the fact that lightning (proved by Franklin to be electrical) was known to leave steel railings magnetised if it struck them. The breakthrough came in 1819 with a discovery by the Danish scientist Hans Christian Oersted. He noticed that a compass needle on his lecture table moved slightly when an electric circuit nearby was turned on. Oersted tried holding the wire above and below the compass and decided that there was an 'electric conflict' running in circles around the wire. We now say that *the lines of magnetic force are in circles around the wire when a current is flowing.*

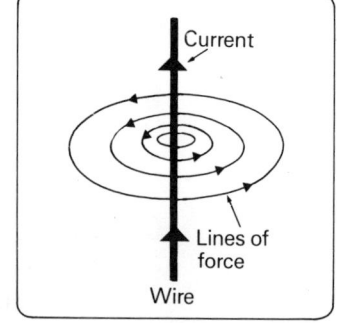

In 1824 William Sturgeon coated a horse-shoe shaped bar of iron with varnish to insulate it, then wrapped a spiral coil of bare copper wire around it so that the turns did not touch one another and the varnish stopped the iron short-circuiting the coil. He found that the iron formed a strong magnet when he connected the ends of the wire to a voltaic pile (battery). *The iron immediately lost its magnetism when the battery was disconnected.*

(A steel bar treated in this way stays magnetised, forming a permanent magnet.) In 1831 Joseph Henry had the idea of wrapping silk thread around wire to insulate it. He could then wrap many layers of wire around the iron bar to make a much more powerful magnet. The electromagnet then made possible the electric telegraph, so that the Morse Code could be used to send messages at incredible speeds over distances of thousands of miles – even under the ocean by cables. The revolution that led to TV and the micro computer had begun!

The investigations of the pioneers are easily repeated in this modern age when we can buy ready-made batteries, compasses and reels of insulated wire.

ACTIVITIES

Work through the investigations below. Note that you are making observations rather than measurements.

SAFETY

In each investigation use only about 3V to avoid overheating.

In each case keep the current on for a few seconds only, to avoid damage to equipment.

ACTIVITY 1   Field around Wire

a Set up the apparatus shown using a low voltage, such as 3V; and 1 m of copper wire, insulated (24 SWG is suitable). Move compass to various positions around the wire. It always shows the direction of the lines of force.

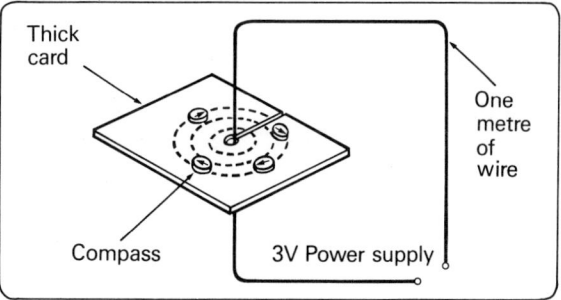

Thick card

One metre of wire

Compass

3V Power supply

b Copy the diagram and state the conclusions to be drawn from your observations.

ACTIVITY II   Field due to Coil

a   Set up the apparatus shown. The long coil is called a solenoid. Connect it to a low voltage, such as 3V.

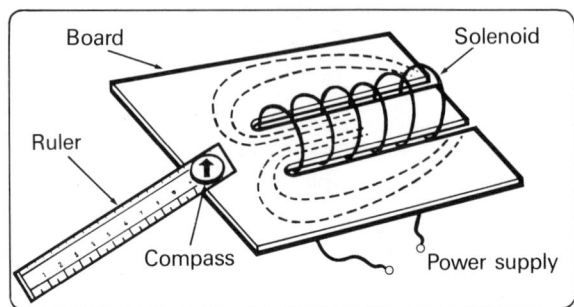

b   Investigate the lines of force with a compass held on a ruler.

c   Also sprinkle filings inside the coil. Gently tap the board to help them line up. (The lines of force do not stop. They go in straight lines inside the coil.)

d   Make a copy of the diagram and state conclusions to be drawn from your observations.

ACTIVITY III   Electromagnet

a   Make a coil of about 20 turns by wrapping copper wire around a former such as a pencil.

b   Will the coil pick up iron filings when the current flows? What happens when the current is switched off?

c   Test the effect on the lifting power of the electromagnet of putting a nail inside the coil to form an iron core.

d   State your conclusions.

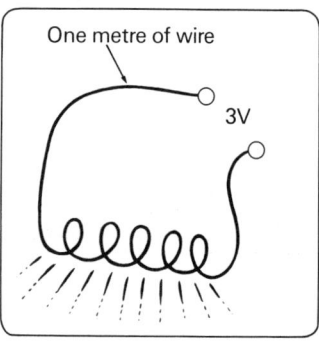

ACTIVITY IV   Magnetic Circuit

The strength of an electromagnet increases greatly if there is an unbroken loop of iron, a magnetic circuit, for the lines of force to pass through.

a   Test this idea by using iron C-cores as shown. Does a piece of cardboard between the cores affect the pulling strength? (The card makes a break in the magnetic circuit).

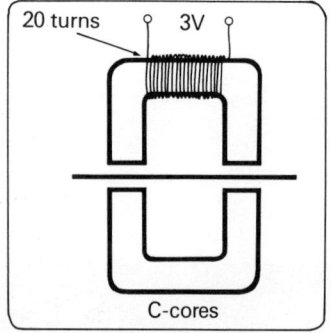

b Copy the diagram and state your conclusions.

ACTIVITY V Field due to C-Core

a Use iron filings and a sheet of paper to investigate the magnetic field due to a C-core.

b How can you tell there is no field when there is no current?

c Sketch the field pattern.

Questions

1 Summarise your observations and conclusions.

2 It is said that lines of force start at the north pole of a magnet and end on a south pole. How did your observations with a solenoid modify this statement?

3 According to the model atom, there are electrons moving in circles around the nucleus of every atom.

    a Why would you expect any atom to be a small magnet?

    b Suggest why some atoms might be very strongly magnetic while others are very weakly so.

4 **Investigation** The strength of an electromagnet depends on the current and the number of turns of wire in the coil. Design an experiment to investigate the effect of each of these variables.

    Planning **a** How will you make it a controlled experiment?

    b What measurements will you take to test the magnetic strength?

    c How can you best display your results?

5 This diagram appears in many physics books to explain the 'clenched fist rule'. What is the purpose of the rule?

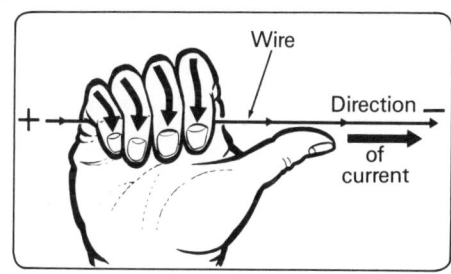

6 A compass needle is placed between two magnets of equal strength, as shown. In which direction will it point?

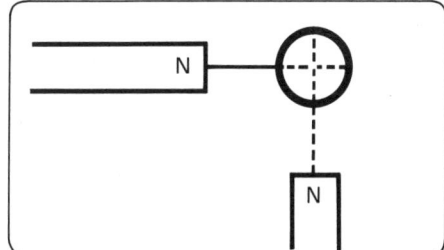

7   Design   The diagram illustrates the principle of instruments used by researchers like Faraday to measure electric current.

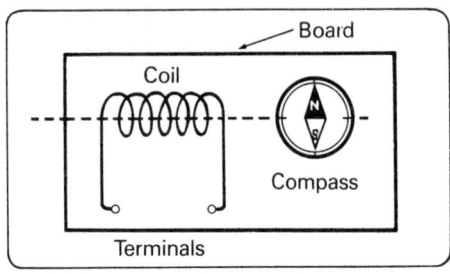

a   Copy the diagram and explain how this 'galvanometer' works.

b   If you made one, how would you calibrate it using another ammeter?

c   How would you modify the design to increase its sensitivity?

d   How might such an instrument be used to measure the strength of a magnetic field, by an astronaut on the moon for example?

✳ 8   The diagram shows the principle of the ticker-tape timer. Copy the diagram and explain what will happen if the coil is supplied with:

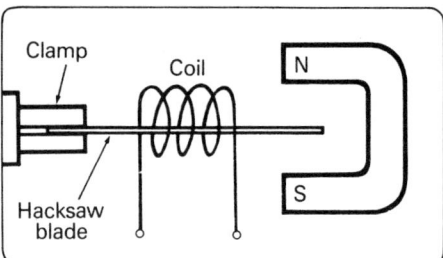

a   DC current

b   AC current.

✳ 9   The diagram shows the principle used in tape recorders to 'wipe' or demagnetise tape. The magnetised steel rod is placed in a coil. AC current is turned on and the rod is then pulled out of the coil. It will be found to be completely demagnetised.

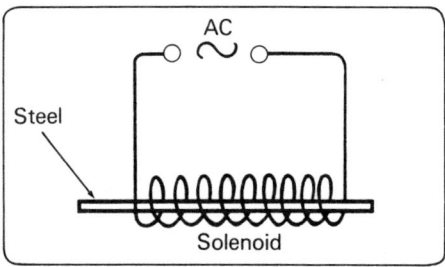

a   Explain how this happens. Begin by considering what happens to the rod before you start to pull it out.

b   Suggest another way of demagnetising without pulling the steel out of the coil.

10   The electric bell makes use of an electromagnet, usually in a horseshoe shape to give two poles and so increase the strength. Copy the diagram.

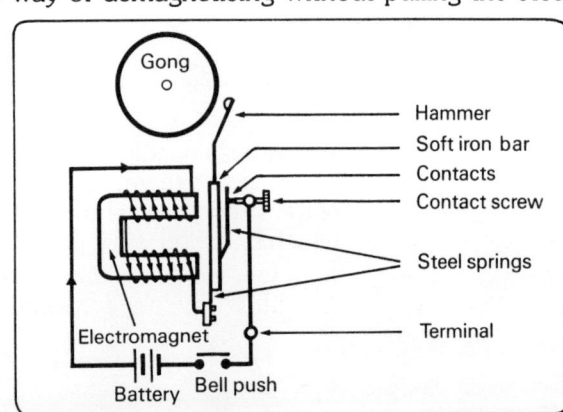

Then copy and complete this explanation of its working.

a When the bell push turns on the current, the . . . is energised.

b The soft iron bar (armature) is attracted to the . . . making the hammer hit the . . .

c When the soft iron moves, the contacts . . . , and the steel spring is . . . .

d With the contacts open, the current is . . . so that the electromagnet . . . the soft iron bar which springs back.

e This turns on the . . . again and the cycle repeats.

11 The diagram shows the construction of a relay. Relays use a small current in a primary circuit to switch on a large or high-voltage current in a secondary circuit. For example a car starter motor draws about 80A of current and so 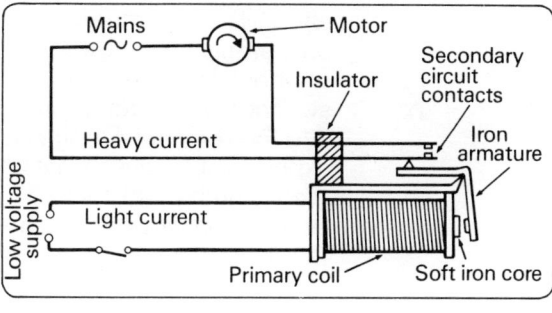 needs very thick connecting wires. This heavy duty circuit is operated via a relay by a small current from the ignition switch.

Copy the diagram and then answer these questions.

a Label the primary and secondary circuits.

b What is the essential device in any relay?

c Why does the armature move when the push button is pressed?

d Why do the secondary circuit contacts close?

e What is the safety function of the insulator pad?

f What should be added to the diagram to make sure the contacts open when the primary circuit is switched off?

g How is the magnetic circuit changed once the relay operates?

h Why does it take more current (pull-on current) to pull the armature on, than it takes (drop-off current) to keep it on?

i How could you measure the pull-on and the drop-off current of a relay?

j What problems would you expect if a relay was left on for a long period of time?

k How would a relay designed to operate on a very small current differ from one for a larger current?

12 In the Faraday laboratory in London you can see the apparatus shown, a compass needle with two wires AB and AC. A, B, and C are terminals. The apparatus was used to demonstrate Oersted's discovery. How did it work?

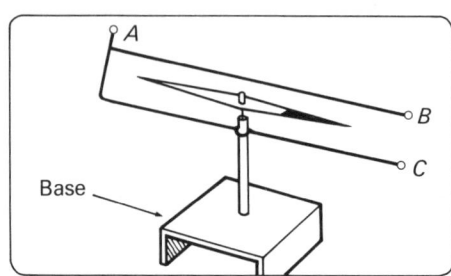

13 **Investigation** Many door bells operate on the 'solenoid effect'. The iron rod shown is only part way into the solenoid. When the current is switched on the rod is pulled into the solenoid. Design an experiment to investigate how the strength of this pull depends on variables such as the current or the length or thickness of the iron rod.

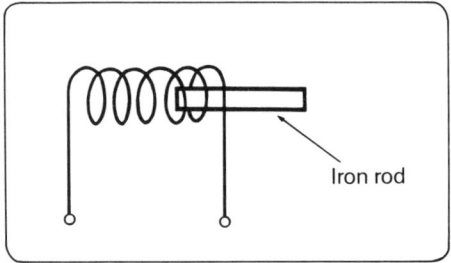

14 **Project a** Make a Morse receiver such as that shown. How will you make it most sensitive?

b How could you use two receivers and two switches to send messages back and forth over one pair of wires?

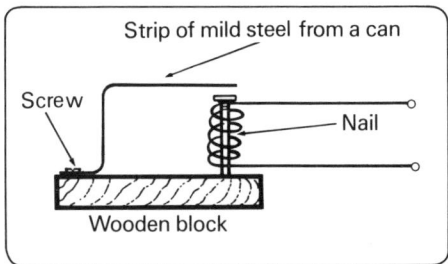

15 **Library** Find out about the international Morse Code.

14   THE ELECTRIC MOTOR

When Oersted made a compass needle move by the magnetic field due to a wire, he had in fact produced a very crude electric motor. In

1821 Michael Faraday succeeded in producing continuous motion of a wire around a magnet, as shown.

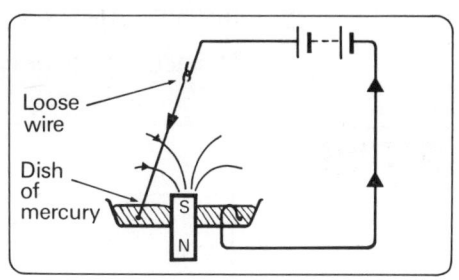

The lower end of the wire moved in a circle around the magnet and kept the circuit complete by touching the mercury.

The direction of the force acting on the wire may be predicted by Fleming's left-hand rule. Holding the left hand as shown, if the First finger points along the magnetic Field, and

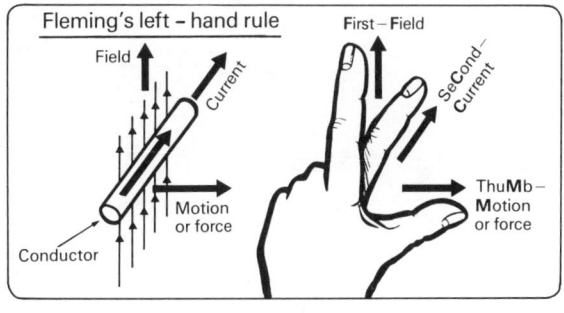

the seCond in the direction of the Current, then the thuMb gives the direction of the Motion or force. Applying the rule to Faraday's motor, the wire would move out of the page at right angles to both current and magnet. In fact, the wire would keep moving in a circle. Reversing the current or the pole of the magnet would reverse the motion.

There are several types of motor, but the modern motor as used in electric drills, starter motors, vacuum cleaners and so on, works as shown in the diagram.

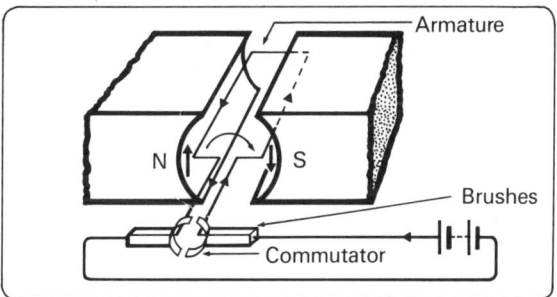

Current enters the armature via the brushes which make moving contact with the commutator. The armature has many turns of wire plus an iron core. By applying Fleming's left-hand rule, it can be seen that the coil will rotate as shown.

The force on the left side of the coil is upwards, and that on the right side is down. When the coil is vertical the commutator reverses the current connections so the forces reverse, keeping the coil rotating. The action of a motor can be studied by assembling a model kit.

ACTIVITY I

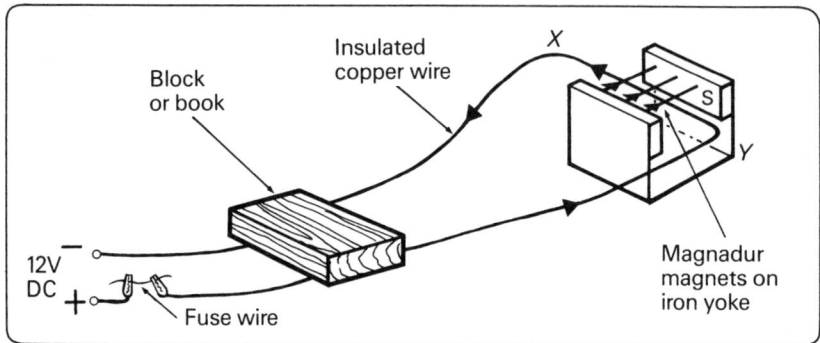

Step 1

a   Set up the apparatus as shown, so that a straight length of wire (XY) sits between the magnetic poles.

b   Using Fleming's left-hand rule, predict what will happen to the wire when the current is switched on. Then test your prediction.

c   Repeat with magnet polarity reversed.

d   Repeat with current reversed.

e   Copy the diagram and record your observations.

Step 2

Predict, then test, what will happen in the situation shown.

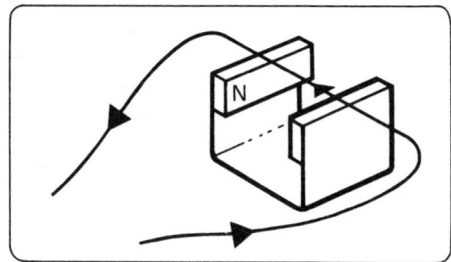

ACTIVITY II

a   Set up the apparatus shown, using an alternating current (AC) supply. The lamp limits the current and prevents overheating.

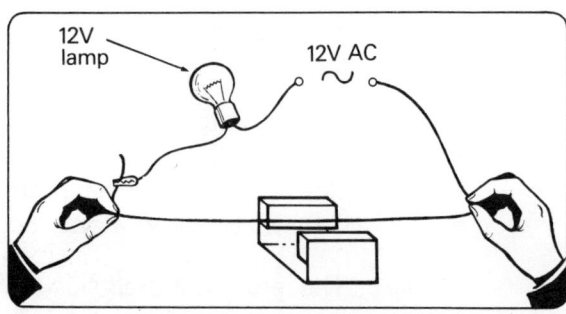

b Pull the wire fairly taut then slowly ease off until a strong vibration occurs.

c Record your observations and explain them.

ACTIVITY III

a Assemble a model kit according to instructions or examine a ready made one. Identify armature, commutator and the brushes.

b Get the motor running on the lowest possible DC voltage. Find the answer to these questions.

   1 For what armature position is there no current in it?

   2 When is the force on the armature greatest?

   3 When is there no force on the armature?

   4 How can you reverse the direction of rotation? (Two methods.)

   5 What factors control the speed?

Questions

1 When current flows, the armature becomes an . . . . One face is a . . . , and is . . . by the north pole of the permanent magnet.

2 There is no current in the armature when it is at . . . to the lines of force.

3 The motor converts . . . energy into . . . .

4 What should the armature coil be wound on for the best results?

5 What factors control the power of the motor?

6 Which parts of a motor wear quickest?

7 The diagram shows the mechanism of a moving coil ammeter. The current enters and leaves the coil by two hair springs. Explain how the meter works. How could the design be modified to make it more sensitive?

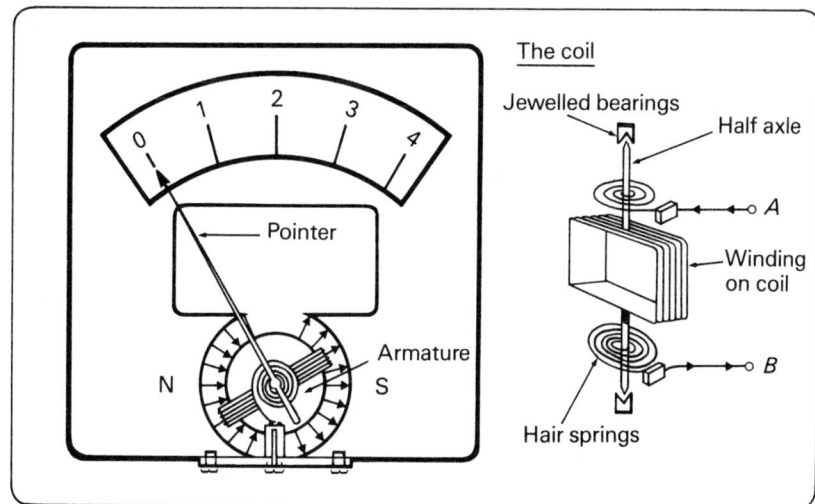

8   Investigation   Plan an investigation using the apparatus in Activity I to find out how the force on a wire in a magnetic field depends on the current.

15   THE DIODE RECTIFIER

A diode rectifier is a one-way valve for electricity. The symbol is as shown:

Notice the arrowhead. Current will only flow through the diode in that direction.

Batteries produce direct current (DC) so that, with a battery, current will flow all the time through the diode or not at all. However, *the mains produce alternating current (AC), in which the current flows back and forth reversing 50 times each second, a frequency of 50 hertz (Hz).* It reaches a peak value in each direction of 340V.

Mains AC voltage varies with the time as shown in the graph, which shows only two cycles.

Much electronic equipment needs low voltage DC current and so must contain a transformer to reduce the voltage, and diodes to rectify the current, to change it from AC to DC.

The AC wave form can be monitored on the screen of a cathode ray oscilloscope (CRO) as shown. Two leads or 'probes' should be connected to form a circuit as shown to the Y inputs of a CRO. As the time base moves the spot across the screen at a steady rate, the AC voltage to the Y inputs moves it up and down, to give the AC wave form.

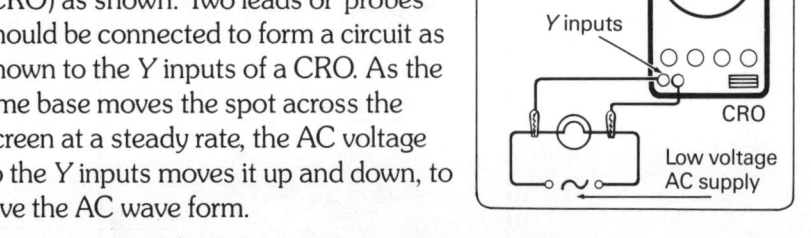

The action of a diode may be investigated as follows.

ACTIVITY I   Diode with DC current

Using the circuit shown, test the effect of reversing the diode.

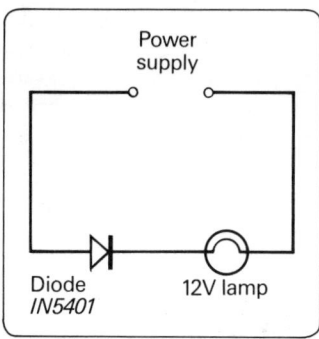

Power supply

Diode
IN5401

12V lamp

ACTIVITY II   Rectifying action with AC current

a Connect the circuit shown using low voltage AC supply. Why does the ammeter show a zero current? Notice the tip of the needle quivering. The lamp shows that current actually is flowing in the circuit.

b Now add a diode in series. Try reversing it. The current is *rectified*. Why is the lamp dimmer?

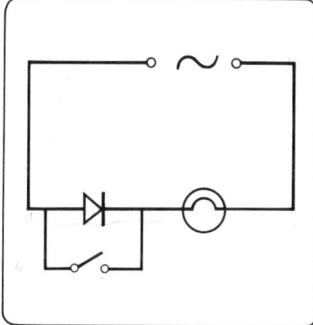

Low voltage
AC

A

c Test and comment on the circuit shown. What useful applications could it have?

ACTIVITY III   Half-wave Rectification

a Connect the circuit as shown. The voltage across the lamp moves the CRO spot vertically as it varies, to give the AC wave form.

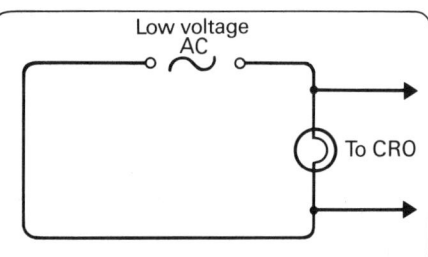

Low voltage
AC

To CRO

b Next include a diode. The CRO should display a half-wave rectified voltage. The diode allows current to flow only half of the time. Reversing the diode will let the other half of the wave pass instead.

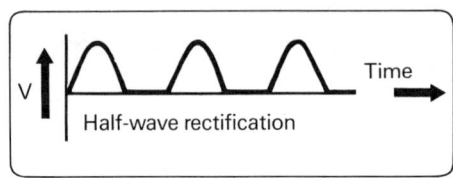

Half-wave rectification

ACTIVITY IV   Full-wave Rectification

Connect up a bridge circuit as shown using four diodes. The CRO should display a full-wave rectified voltage. By tracing the current for each half of the AC cycle you will find that it always flows in the same direction through the lamp. Full-wave rectified current is much more like steady DC current than half-wave is.

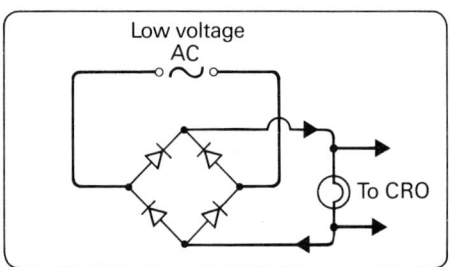

Low voltage AC

To CRO

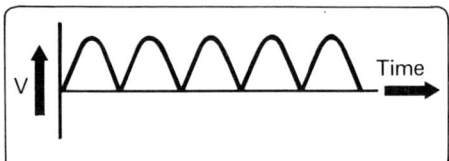

Questions

1 Copy the circuits you used and add notes on each.

2 What do the letters CRO stand for?

3 What does the CRO time base do to the spot?

4 What do the Y inputs do to the spot?

5 Copy this circuit and draw the current pattern at the points *A*, *B*, and *C*.

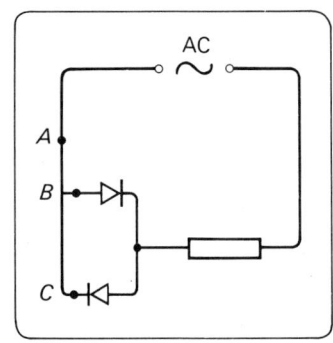

AC

6 Copy and label this battery charger circuit. The transformer reduces the mains voltage to about 20V.

a Explain how the charger works.

b Which is the positive output terminal?

c Where would you add an ammeter?

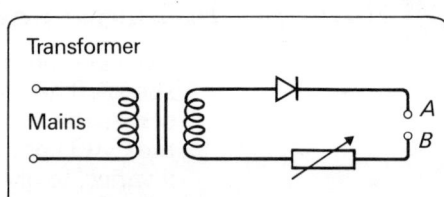

Transformer

Mains

✱ 7 The circuit shown is useful with soldering irons. Make a copy and explain why it is used. How could it be used in kitchen appliances and why?

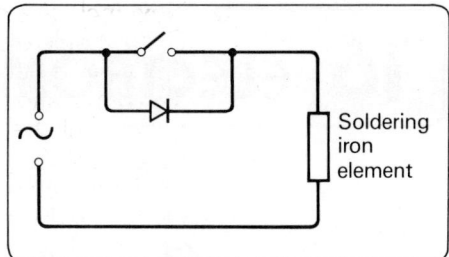

Soldering iron element

8 How could you use a CRO to measure the peak voltage of an AC supply? How would you calibrate it? Would you have the time base on or off?

9 Tell what will happen and will be seen with each circuit (the ammeter is a DC ammeter).

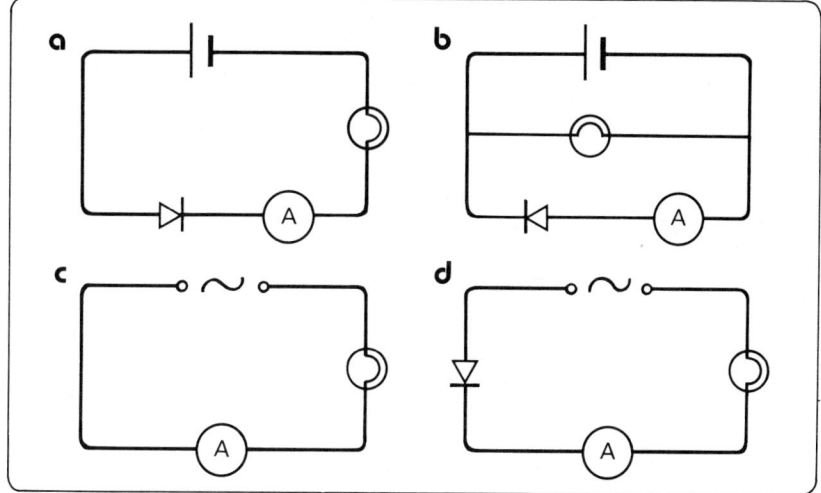

10 **Design** **a** How would the circuit shown be useful on a bike?

b If the battery was rechargeable, how could the dynamo be switched to recharge it?

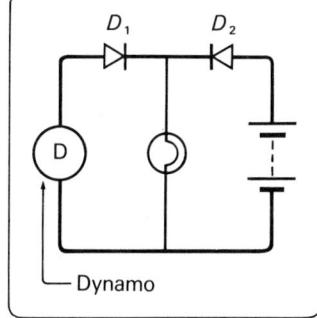

D_1 D_2

D

Dynamo

16 ELECTROMAGNETIC INDUCTION

In 1831 Michael Faraday wrote a letter to a friend saying 'I am busy just now again on electromagnetism, and I think I have got hold of a good thing.' The good thing Faraday had discovered was the secret of using magnets to generate electric currents in wires. This great discovery has made possible the millions of dynamos or generators that produce our mains electricity and that operate the electrical systems of cars, aeroplanes, and motor cycles.

A few years earlier Oersted had discovered how an electric current could produce a magnetic field, leading on to the invention of the electric motor. Faraday's discovery of electromagnetic induction led to the invention of the dynamo, to give the world the powerful dynamo-motor combination that is vital to our technological civilisation.

The dynamo converts mechanical energy to electrical energy. That energy can then be carried by current along wires to a motor which reverses the process, and converts electrical energy into mechanical energy, to operate drills, lathes, and the host of domestic appliances we call 'energy slaves'. In this way energy from a distant power station can be used to do work in our homes and factories.

Faraday's Law of Induction

Faraday investigated induction using apparatus like that shown. When the magnet is pushed into the coil, a current is induced (generated) in the coil, and the needle of the sensitive milliammeter flicks to one side. When the magnet stops,

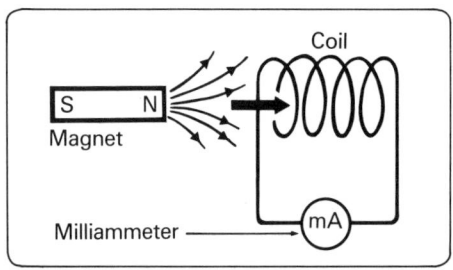

no more current flows. Then, when the magnet is withdrawn, the needle flicks in the other direction. Using a more powerful magnet, moving it faster, and having more turns on the coil all cause a larger current to be induced. And moving the coil instead of the magnet also generates a current. How could this be explained?

Faraday used the lines of magnetic force model (idea) to explain the process in a very simple way. He said that a voltage (e.m.f.) is generated in a wire whenever it cuts through lines of magnetic force. He then summarised his observations in *the famous law of electromagnetic induction: whenever lines of force are cut, an e.m.f. is generated, which is directly proportional to the rate at which the lines are cut.*

We can understand Faraday's law better by seeing how it is applied in the design of the modern dynamo.

The AC Dynamo or Alternator

One type of alternator uses a coil rotating between the poles of a magnet to cut lines of force, as shown.

As the coil (armature) rotates, the two long sides of it cut the lines of force. When it is horizontal as shown, the AC output voltage reaches its peak value as lines of force are being cut at a maximum rate per second. When the coil becomes vertical the output voltage reaches zero as its edges brush along the lines of force, not cutting across them. The voltage then reverses as the coil moves on around, cutting the lines of force in the opposite direction. One rotation of the armature generates one cycle of AC voltage, as shown.

To increase the peak voltage, more lines of force must be cut each second. This can be done by having a stronger magnet, and many turns of wire on the armature. Increasing the speed of rotation will also boost the peak voltage, but will make each AC cycle shorter. The generators that produce our mains supply make 50 cycles per second with a peak voltage of 340V.

The DC Dynamo

When a car is stationary or moving slowly, its electrical system is operated by a 12 volt battery which produces DC current. At higher speed, the dynamo takes over and also recharges the battery. For this reason a DC dynamo output is needed. The modern car uses an alternator and rectifies its output to DC using diode rectifiers.

However, older cars have a DC dynamo which has the same structure as an electric motor. This time the current is drawn from the rotating coil via a commutator instead of slip rings.

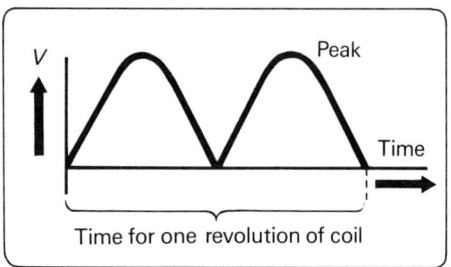

Because the commutator changes contact from one brush to the other just as the AC voltage in the coil reverses, the output voltage never reverses. It is DC, but it is not steady DC as from a battery. It varies from zero to peak value, as shown, like full-wave rectified current.

Because they have the same structure, a DC dynamo can be used as a motor and vice versa. The famous tractor inventor Harry Ferguson suggested building an electric car with a drive motor on each wheel. To slow down the vehicle, the brake pedal would switch off the current. The car's motion would then work the motors as dynamos, using up its kinetic energy to recharge the batteries and rapidly bring it to a halt.

Questions

1   The diagram shows the basic construction of a bicycle dynamo. This type of dynamo was invented by H. Pixii in 1832. (Pixii also invented the split ring commutator used in motors.) It uses a rotating magnet instead of a rotating coil, so that no sliding contacts are needed to draw current from it.

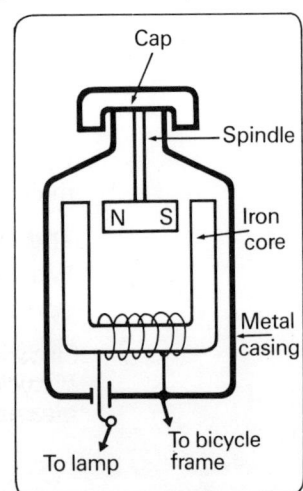

   a   How would you alter the design to generate a higher voltage?

   b   Why might its performance fall off after a few years?

   c   Does it generate AC or DC current?

   d   Why are cycle lights brighter going downhill?

   e   Draw a graph to show how the current from such a dynamo varies with speed. Give as many details as possible.

   f   What energy changes take place?

g What advantages does a dynamo have over a battery? What disadvantage?

h How could a bicycle dynamo be adapted for use as a speedometer?

2 A new teacher found the device shown in an 'energy kit'. It looked like two motors with the spindles joined by valve rubber.

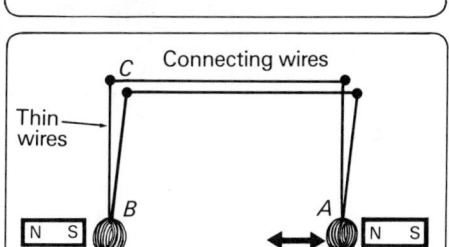

a What was it, and what demonstrations could be done with it?

b Explain in detail how its energy conversion efficiency could be measured.

3 The diagram shows a demonstration of induced currents. If A is set swinging, B will start also. Explain step by step how this happens. How would you try to make the swing as big as possible?

Connecting wires

C

Thin wires

B

A

N S Coil

N S

Coil

✱ **4** **Design** An engineer made a prototype spark device to ignite the oil in a central heating boiler, using the apparatus shown.

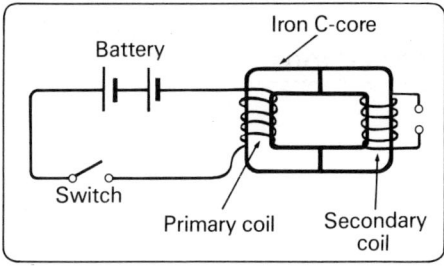

Iron C-core

Battery

Switch

Primary coil

Secondary coil

a Explain how it works.

b How would you modify the design to give a stronger spark?

c How could an electric bell be added to try to produce a continuous spark?

5 Why might tiny electric currents flow in the metal wings of a flying airliner? What would control the size of these currents? How could you test your explanation using a length of wire and a milliammeter?

✱ **6** How might a dynamo be used to attempt to measure the strength of a magnetic field, by an astronaut on the moon for example?

7 **Investigation** How could you use a CRO to investigate the output of a cycle dynamo? Explain fully what you would do and what measurements you would take.

8 When Michael Faraday demonstrated his discovery at the Royal Society, one MP said, 'Yes, Mr Faraday. Very interesting. But what good is it to anybody?' The wise reply was 'What good is a new-born baby, Sir?' What did Faraday mean? Was he correct?

ACTIVITY

Step 1

Set up the apparatus shown using a milliammeter, coils of say 100 turns of wire, and 200 turns, and a weak and a strong magnet.

Find the answers to these questions about the e.m.f. induced when lines of force from the magnet are cut by the wires of the coil.

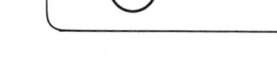

Move magnet

a How does speed of movement of the magnet affect the e.m.f.?

b How can you generate an AC voltage?

c Does moving the coil instead of the magnet work?

d What effect does the polarity of the magnet have?

e Does the number of turns on the coil make any difference?

f What happens if the magnet is moved sideways instead of into the coil?

Step 2

You are given a magnet with no polarity markings. Use the apparatus to determine which is the north pole.

ANALYSIS

1 A voltage (e.m.f.) is . . . when . . . are cut by a wire.

2 The induced current was measured by the . . .

3 The three factors controlling the strength of the induced current were . . .

4 Suppose the milliammeter was disconnected. Would there still be an induced voltage?

5 With the coil, each . . . acts as a tiny . . .

17   THE CAPACITOR

Early experimenters with electricity looked for some way to store it. It seemed obvious, perhaps, to store the electricity in a bottle, like water. In 1745, E. G. von Kleist connected a static electricity generator to a nail held by a cork and dipping into water in a large glass jar. It was found that sparks and shocks could then be obtained by touching the nail, showing that electricity had been stored. The first shocks terrified those who received them. However, it soon became fun and people would line up to form human chains by holding hands. When the first one touched the nail, the shock made everybody jump at the same time. This happened because the muscles of the body are operated by tiny electrical signals from the brain. Shocks can be lethal because they may stop the heart which is also made of muscle.

These first capacitors were called Leyden jars, named after a town in Holland. Ben Franklin charged a Leyden jar from the string of his famous kite and later used a spark from it to light a fire.

The Leyden jar was soon improved: it was found to work better if the inside and outside were coated with metal foil. Contact to the foil inside was made by a chain dangling from a nail held in the cork. Then someone realised that the jar was simply a sheet of glass (shaped like a bottle) sandwiched between two metal sheets, an insulator between two conductors. So the next step was to use a large sheet of glass between sheets of metal foil. The larger the sheets the greater the storage capacity, and hence the name 'capacitor'. Making the glass thinner also increased the capacity. The capacitance is directly proportional to the surface area of the plate and inversely proportional to the distance apart.

Modern capacitors have incredibly thin insulating layers between sheets of foil. The sheets are then rolled up into tubular shape to reduce the bulk of the capacitor.

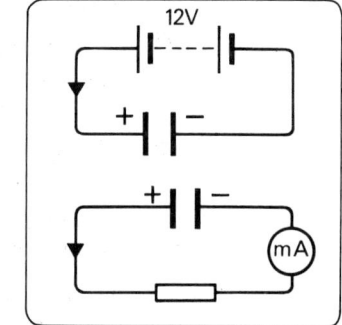

The diagram shows a capacitor connected to a 12V battery. Current flows into the capacitor until it is fully charged. If the capacitor is then connected to a resistor, it will act like a 12V battery and drive current through the resistor. However, as it gives out charge, its voltage rapidly falls and it soon becomes discharged.

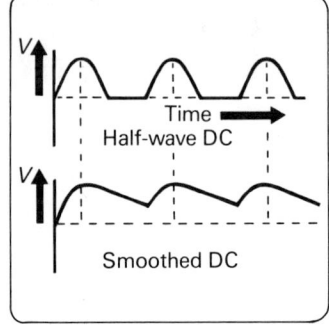

Capacitance is measured in farads. Because the farad is so large, capacitors are used marked with values in microfarads (millionths of a farad), µF, or even picofarads (thousandths of a microfarad), pF.

This storage function of capacitors is used to smooth half- or full-wave rectified current, as shown.

The capacitor charges up to the peak voltage then discharges to fill the gaps when the supply voltage falls. The larger the capacitor, the better the smoothing effect.

The circuit diagram shows how the smoothing capacitor is connected to the output of a power supply.

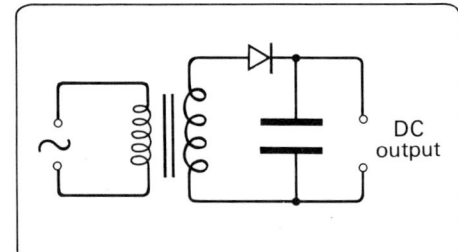

Questions

1 Draw a diagram to show the structure of a metal foil Leyden jar.

2 What factors control the storage capacity of a capacitor?

3 Copy the circuit for the smoothed DC supply and explain how it works.

4 A capacitor connected to a voltmeter will slowly discharge through the voltmeter. Sketch a graph to show how you think the voltmeter reading would vary with time.

5 **Design** How would you try to construct a capacitor using materials found in the kitchen?

6 Using the water model of electricity, the capacitor can be compared to a water storage tank, as shown. The pump acts like a battery.

   a Assuming the tank is tall, what will happen after the pump has been running for a few minutes?

b How could the tank be used to compare the pumping pressure of two pumps?

c Suppose the pump was removed and the water allowed to run out. Sketch a graph to show how the flow rate would vary with time.

d How would a tank that represented a smaller capacitor be different?

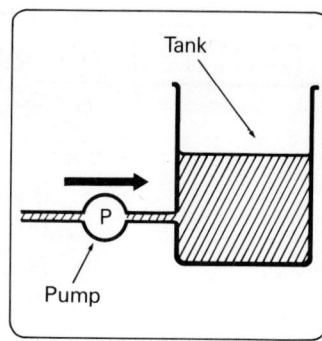

7 Why are capacitors not used as rechargeable batteries?

8 On a mains-operated radio, the sound continues for a short time after switching off. Why is this? Why does this not happen with a battery-operated radio?

✱ 9 **Design**   You are given some 1000μF capacitors and a 9V battery. Design a switching circuit to allow the capacitors to operate a 90V camera flash. How could you maximise the energy output per flash?

ACTIVITIES

The storage capacity of a capacitor is given in microfarads. Examine several capacitors and note their values.

The thin insulating layer will break down if too high a voltage is applied. Note the voltage rating of each capacitor.

Some capacitors are of the electrolytic type and will be destroyed if connected wrongly. The end marked + must always be connected to the positive lead.

ACTIVITY I   Storing Charge

a Charge a very large capacitor (such as 2200μF) on a 9V battery by touching it to the terminals.

b Now test it with a voltmeter. Notice how it loses its charge.

c Try smaller capacitors. How long does each take to discharge fully? Display your results in a table then plot a graph of discharge time in seconds against capacity in microfarads.

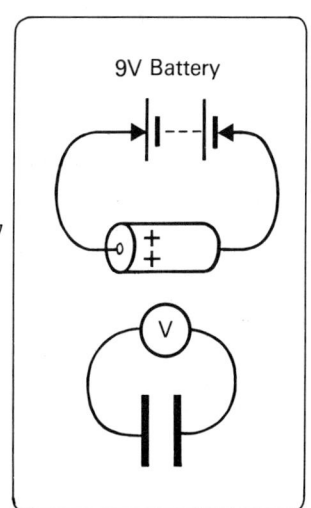

ACTIVITY II   Smoothing Capacitor

a   Test the circuit shown *without* the smoothing capacitor. CRO will show half-wave rectified current, unsmoothed.

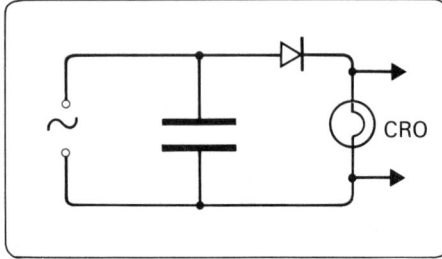

b   Now add a capacitor. It should discharge in gaps to help smooth current flow. Compare the smoothing effect of several capacitors.

Unsmoothed   Smoothed

18   THE TRANSFORMER

The original domestic electricity supply was DC. The problem was that the resistance of the cables reduced the voltage, so that people far away from the power station had a lower voltage than those nearby. This problem was solved by using AC current so that a transformer could be used to adjust the voltage at local substations. Transformers do not work on DC voltages, only AC.

Today the National Grid system uses 400 000 volts on its main lines because transmission at high voltage and low current is much more efficient than at low voltage and high current. At the consumer end, step-down transformers are used to reduce the voltage at local sub-stations. It is reduced to 132 000V for heavy industry, 25 000V for trains, 11 000V for small factories, and 240V for homes.

The diagram shows the structure of a step-up transformer (one used to increase voltage). It consists of two coils wound on a core made of layers of soft iron. The AC current into the primary coil turns the core into an AC electromagnet. Lines of force shoot back and forth through the secondary 50 times a second and induce an AC voltage in each turn of the wire in it. *By having more turns on the secondary, a greater output voltage is*

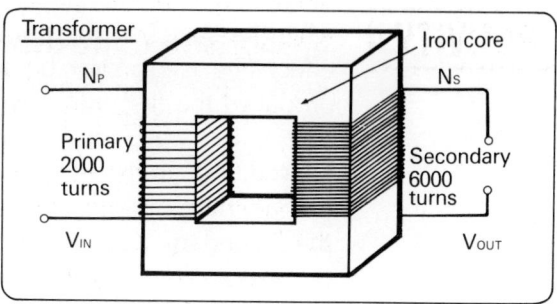

Transformer

N_P

N_S

Iron core

Primary 2000 turns

Secondary 6000 turns

V_{IN}

V_{OUT}

produced, and the voltage is stepped up. By having fewer turns on the secondary than on the primary, the voltage is stepped down.

In the diagram N_P is the number of turns of wire on the primary coil, and N_S, those on the secondary coil. V_{IN} is the input voltage; V_{OUT} is the output voltage.

The voltages are related by this formula: $\dfrac{V_{OUT}}{V_{IN}} = \dfrac{N_S}{N_P}$

Example: For the transformer shown, the turns ratio $\dfrac{N_S}{N_P} = 3$. If used on the mains voltage (240V) the output, V_{OUT}, would be 3×240 or 720V. Step-up models like this are used to run neon signs.

If the situation was reversed with 6000 turns on the primary and 2000 on the secondary, the voltage of the mains would be reduced to 80V, 240V divided by 3.

As well as inducing currents in the secondary coil, the lines of force will induce them in the iron core. To reduce the energy loss and heating due to these 'eddy currents', the core is made of thin layers of wire separated by coatings of insulation.

For a perfectly efficient transformer, the power output ($I_{OUT}.V_{OUT}$) would equal the power input ($I_{IN}.V_{IN}$). In any case, output power cannot exceed input. This means that a step-up transformer gives a higher voltage but a lower current, and vice versa.

Example: Find the current drawn from the mains when a 12V model car transformer supplies a current of 5A.

Working: $I_{IN}.V_{IN} = I_{OUT}.V_{OUT}.$ $\therefore 240.I = 5 \times 12$
$\therefore I = 0.25A.$

Michael Faraday experimented with the first transformer by having two coils of wire wound onto a wooden rod. He hoped that a current would flow in the secondary coil when one flowed in the primary. This did not happen, but one day he noticed a flicker on the galvanometer when he switched the primary off. He then found a much greater effect if he used an iron bar in place of the wood. His next step, which produced the first transformer, was to wind both coils onto an iron ring. In this way most of the lines of force produced by the primary passed through the secondary. So that switching the primary on and off had the same effect as pushing and pulling a magnet in and out of the secondary. Later, when dynamos like that of Pixii were built, AC voltage could be applied to the primary. This work opened the door to our modern power system.

The National Grid

Electric power is distributed across the nation using a network of cables called the National Grid. In transmitting power along wires, we have the choice of low current-high voltage (IV) or high current-low voltage for the same amount of power. Because the power loss in a given wire increases with the square of the current, it is more economical to make V very large and I as small as necessary. As a result voltages on cables in the supergrid sections are 400 000 volts.

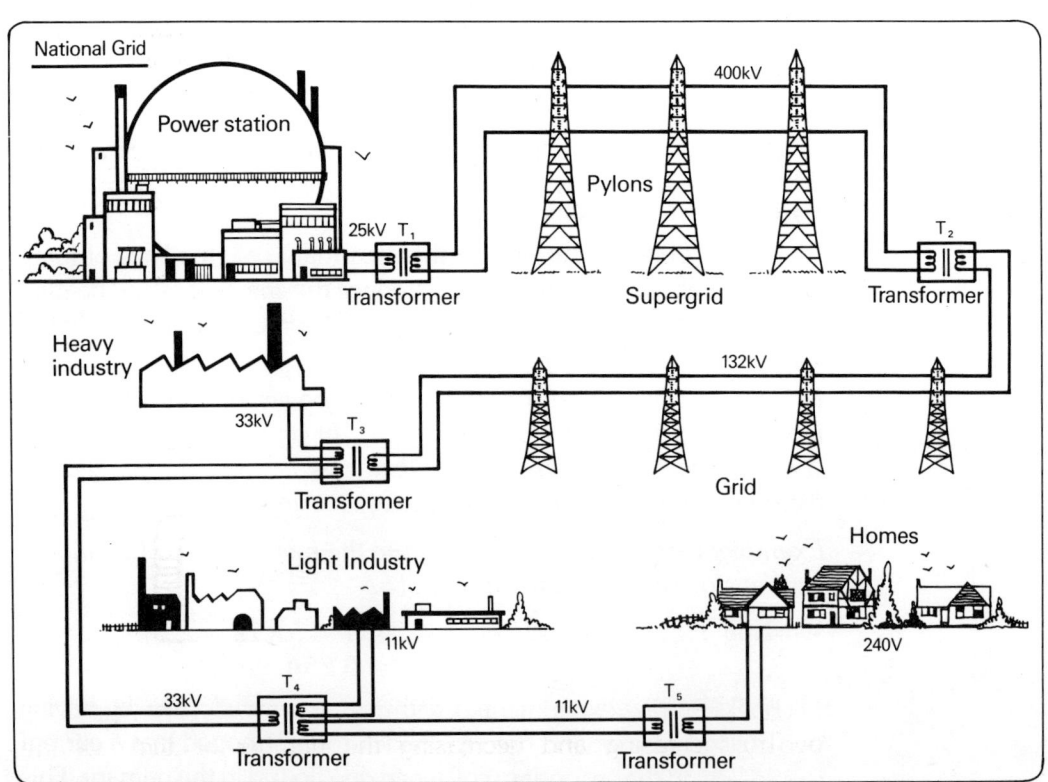

The principles of the transformer may be investigated as follows.

ACTIVITY I

Michael Faraday used apparatus like that shown.

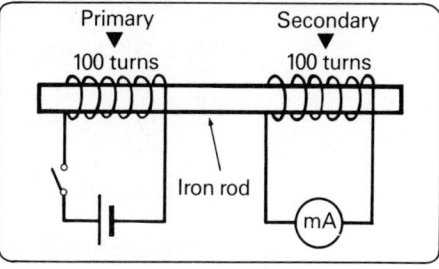

a Look for induced current in the secondary when the primary is switched on or off. This is the only time the lines of force are moving and being cut.

b Can you generate a kind of AC current by rapidly switching the primary on and off?

c Copy the diagram and state your observations.

ACTIVITY II

a We can produce rapidly moving lines of force by applying AC voltage to the primary.

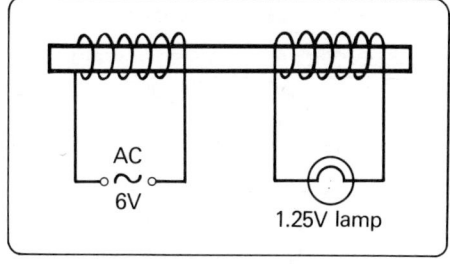

b This is a transformer. Does it work better if the coils are pushed closer together so that more lines of force go through the secondary coil?

c Copy the diagram and record your observations.

ACTIVITY III

a The transformer works much more efficiently using an unbroken iron core (two C-cores) as shown. The cores should fit together well, because a small air gap greatly reduces the magnetic field in them.

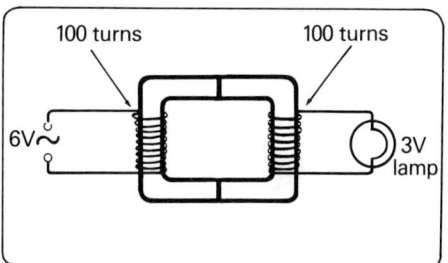

b Try increasing and decreasing the number of turns of the secondary.

Questions

1 a Copy this circuit symbol for the transformer. What do its three parts represent?

   Copy and complete the following two sentences.

   b Transformers only work on . . . voltages.

   c The iron is laminated in order to . . .

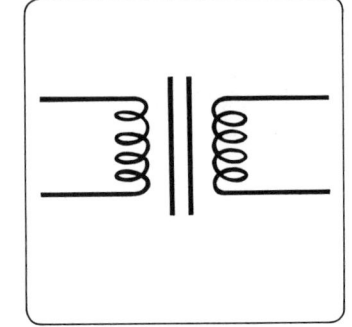

2 a A transformer has 2000 turns on the primary and 1000 on the secondary. A voltage of 24V is applied to it. What will it do to the voltage?

   b A transformer is being designed to step down the mains voltage to 12V to run a slot-car set. Suggest the number of windings needed.

   c In the last question, why might you need a few more secondary turns than you suggested?

3 How would you design the transformer to run a 1200V neon sign?

✱ 4 In a spot welder, two sheets of metal are gripped between two pointed jaws which are connected to the secondary of a step-down transformer. A large current flows through the jaws and melts the metals so they weld. Suppose the welder works off a 240V supply and has a 13A fuse in its plug. If the secondary voltage is 40V, find

   a the turns ratio $\dfrac{N_S}{N_P}$ of the transformer

   b the maximum current that can be produced

   c the power output being used to melt the metals.

5 **Investigation**   Using two coils on an iron bar as a transformer, does the output voltage depend on the distance between the two coils? Design an experiment to investigate this variable. The results should be presented as a graph. What safety precautions would be taken?

6 The following data was obtained with a model transformer. Analyse the data graphically and state what conclusions can be drawn from it.

N_P	N_S	V_P	V_S
500	125	12	2.9
500	250	12	5.8
500	500	12	11.7
500	2000	12	47.5
500	2500	12	59.3
500	3000	12	71.5

7 A battery charger produces an output of 16V and charges a car battery at a current of 2A.

  a Calculate the turns ratio of the transformer.

  b Calculate the power output of the charger.

  c Assuming 100% efficiency, calculate the current drawn from the mains.

✱ 8 **a** Copy the National Grid diagram and work out the turns ratio for each transformer saying whether it is a step-up or step-down.

  b If a consumer switches on a stove and other appliances to use 40A of current, calculate the extra current in the supergrid as a result.

  c If it takes about 20 000V to create a spark one centimetre long in air, roughly how far could a spark jump from a 400 000V cable?

19 THE DOMESTIC ELECTRICITY SUPPLY

An American fan heater would overheat and burn out if used in the United Kingdom. This is because the mains voltages are different in the two countries: 115V in America, 240V in Britain. Since power consumption depends on the square of the voltage, the heater element would produce about four times as much heat.

In all countries, the mains supply is AC (alternating current) not DC (direct current). The diagram shows one cycle of UK mains supply voltage. Notice that *the voltage reaches a peak value of 340V in each direction. This voltage applied to a resistance wire would produce the same heating effect as a steady* 240V DC. *This effective mains voltage is called the RMS (root mean square) voltage, obtained by dividing the peak value by* $\sqrt{2}$. It should be clear that mains electricity is far more dangerous than the RMS figure suggests. Because of the lethal voltage involved plus the hazard of fire due to short circuits from faulty work, it is extremely unwise to do any of your own house wiring.

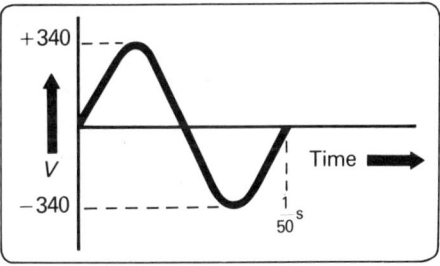

All the wiring in a house leads back to the consumer unit ('fuse box') which connects it up to the electricity board supply. From the consumer unit, cables run to wall sockets and lights. Equipment that draws very large currents, such as stoves, immersion heaters and electric showers, has its individual cable connecting it to its own fuse or circuit breaker in the consumer unit. The thickness of the wire in the cables depends on the current it has to carry. A wire carrying 30A to an electric shower will generate about 36 times as much heat as one carrying 5A for a lighting circuit. So, using wire that is too thin could result in a fire. On the other hand, thicker cable is much more expensive.

Lighting Circuits

Lights are connected to the consumer unit by two cables, live and neutral. *The neutral stays at zero voltage while the live varies from +340 to −340 volts, producing AC current.* The diagram shows a typical lighting circuit. The circuit is protected by a 5A fuse, which allows about a dozen 100W lamps to be on at one time. Notice that *the fuse goes in the live wire so that all parts of the circuit are dead if the fuse blows* (or circuit breaker trips).

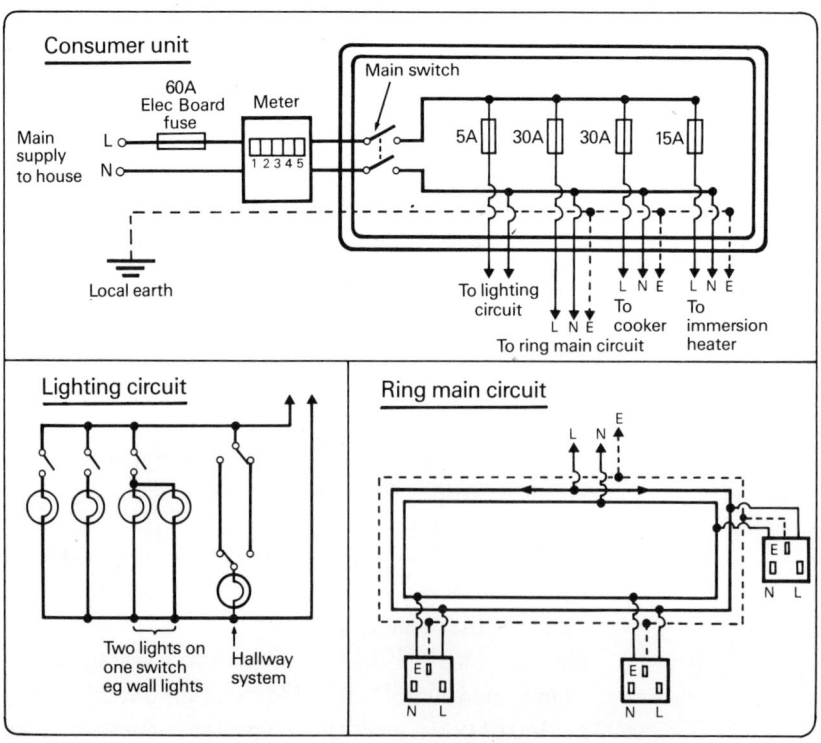

Power Circuits

Cables from the consumer unit to power sockets contain a third wire, the earth, as well as the live and neutral. The earth is a safety feature. A thick wire connects the earth terminal of the consumer unit to a metal plate buried in the ground outside the house. At one time metal pipes could be used to make connection to earth, but with the use of plastic pipes, a direct earth must often be set up.

If there is a fault in an appliance and the live wire contacts the metal frame, current will flow to earth instead of to the neutral wire. If there was no earth connection, current could flow through a person touching the appliance, perhaps killing them. Because lighting circuits have no earth wire, it is unwise to connect any appliance to a light socket. There was a tragic case of a small boy being killed when he picked up a faulty inspection lamp his father had connected to the lighting circuit in the garage. It should have been fitted to a fused plug with an earth.

Sockets are connected to the consumer unit by a ring main circuit as shown. *Typically a ring main runs around a roof space. This makes it very convenient to add extra sockets when needed. Notice that the current splits up and flows two ways around the ring. This reduces the heating effect and enables thinner cable to be used. Typically a ring main is controlled by a 30A breaker.*

Questions

1   Copy and complete the following sentences.

   a   The letters AC stand for . . . .

   b   The letters DC stand for . . . .

   c   In the United Kingdom, the peak voltage of the mains is . . . .

   d   The average or RMS voltage is . . . .

   e   All wiring in a house leads back to the . . . .

   f   Lighting circuits have a . . . fuse, but power circuits have . . . .

   g   The . . . wire is at zero voltage, while the live wire . . . .

   h   If the live comes in contact with the metal casing of an appliance, current will flow . . . .

   i   Lighting circuits have no . . . wire because . . . .

   j   An electric shower is not part of the ring main circuit because . . . .

2 a A 200W lamp takes about 0.8A. How many such lamps could be switched on at once in a typical house circuit?

   b A fan heater draws about 12A. How many could run off a ring main at one time?

   c What wiring is protected by the fuse in the plug of an appliance?

   d What current flows in a ring main cable when a fan heater is turned on?

   e What wiring is protected by the fuses/circuit-breakers in the consumer unit?

3 a What are the advantages of a ring main?

   b Why is a fuse located in the live wire?

   c Why is cable for electric showers very expensive?

   d What would you expect to happen if a British kettle was used in America?

   e In America the mains voltage is 115V. Calculate the peak voltage.

4 Design How could you make a model using materials such as wood, nails, wire, etc. with a power supply to demonstrate the principles of the ring main to a class?

5 If a 1kW appliance draws about 4A find the current taken by the following appliances:

   a 8kW shower

   b 3.5kW immersion heater

   c 10kW cooker.

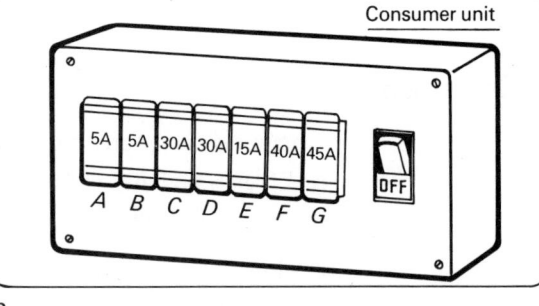

6 Use the answers to the previous questions to help identify the fuses (or breakers) in the consumer unit diagram.

 **ACTIVITY**

NOTE: This should be done under close supervision of teacher or parents.

Step 1

   a Examine an appliance or item of equipment and make a note of its stated power rating (wattage).

b Open the plug and make a note of the rating of the fuse.

c Appliances under 700W should have a 3A fuse or smaller. Those over 700W need a 13A fuse. Comment on the suitability of the fuse.

d While the plug is open, check on the tightness of the screws. Make a note of any needing attention.

e Repeat for other appliances.

Step 2

Draw up an inspection report and work out the percentage of wrongly fused and faulty plugs.

20   ELECTRICAL SAFETY

The human body works electronically. Tiny currents flowing along nerve fibres from the brain cause muscles to contract. By means of opposed muscles the joints of the body can be moved back and forth. For example, one set of muscles is used to clench the fingers into a fist. Another set is used to unclench them. *An electric shock causes currents to flow in the body which is a very good conductor beneath the skin.* These currents can operate the muscles so strongly that opposing signals from the brain have no effect. For example, if you touch a live wire your hand may involuntarily close around it and nothing you can do will make the hand let go. A few milliamps of current is enough. Your throat may also be paralysed so that you cannot shout for help. Meanwhile the currents flowing through the flesh can produce severe, perhaps lethal, burns.

The heart is composed of muscle tissue and operates electrically. This is why some people with unreliable heart action have pacemakers fitted which send impulses to the heart at regular intervals. *It is not surprising then that electric shocks can stop the heart – death by electrocution,* as in the electric chair.

Shocks will be much more severe if your body is well earthed. For example, if you were touching a metal sink unit with one hand and made electrical contact with the other, current would flow up one arm, through the chest and down the other arm.

Electricity and water do not mix! This is a good safety motto.

Although pure water is a poor conductor, the addition of salt from your skin makes it a good conductor. This is why wet hands should not be used with electrical appliances. And nobody with any sense would think of taking a radio or fan heater on an extension cable into a bathroom. By law, electric sockets must not be fitted in bathrooms or within reach of a kitchen sink unit (because of the water and because it is such a good path to earth). Lights in bathrooms should operate by pull switches on the end of long insulating cords. A 'cowboy' electrician may be willing to ignore such regulations, but they are there for your protection.

Modern sockets automatically shutter off the live and neutral holes so children cannot poke metal objects into them. You will notice that the pins on some plugs are partly sheathed in plastic so that a person grasping the plug carelessly will not accidentally touch a live pin as it enters or leaves the socket.

As explained earlier, the earth wire connects to the appliance frame to protect the user in case the live wire makes contact, perhaps because of faulty insulation. *It is important for the earth wire to be tightly attached inside the plug of the appliance. As a safety routine, all plugs in homes and the workplace should be checked at least once a year.* At the same time the live and neutral terminals in each plug should be checked. A loose connection has a high resistance and can cause the plug to heat up and is a potential fire hazard that fuses cannot protect against.

When using electric hedge trimmers and lawn-mowers, you face the hazard of cutting the supply cable. This can be lethal, especially as the first reaction is often to bend down and pick up the severed cable to examine it! *It is a wise precaution to fit a residual current device which switches off the mains in a fraction of a second when things go wrong.* Normally the currents in the live and neutral wires will be exactly the same, although in opposite directions. The residual current device monitors these currents and switches the supply off if they differ by as little as 1 milliamp.

Likewise a current in the earth wire means that something is wrong. An earth-leakage trip monitors the earth wire and switches off the supply when a tiny current flows. A fuse is not enough protection. You can be killed by a current of only 0.1A, far too small even to begin to blow a fuse.

Questions

1 Volta and Galvani discovered how to make batteries after noticing that frogs' legs twitched when in contact with metal instruments in salt solution. Explain what this statement has to do with electrical safety.

2 Why do ambulance workers sometimes use high voltage electrodes on people who have been injured?

✳ 3 a Suppose a person standing on a wooden floor offers a resistance of 40kΩ from finger tips down to earth. He then touches the live wire just at its peak. Calculate the current that will flow through him.

b If a current of 10mA causes involuntary muscle spasm, would you classify the shock of the person as mild, severe, or fatal?

c Suppose the accident occurred while standing on a damp lawn so that the body-to-earth resistance was only 10kΩ. Comment on this situation.

✳ 4 The diagram shows the principle of a residual current circuit breaker. On the principle of the transformer, the AC current in live wire L will induce a voltage in secondary coil XY.

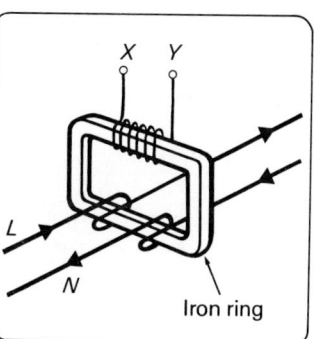

a Why then will there normally be no voltage in coil XY?

b Under what circumstance will there be a voltage induced in XY?

c Suggest ways to increase the sensitivity of the device, so that a larger voltage is induced in XY.

d How could a relay be added to the device?

5 Suggest some potentially dangerous situations with electricity and how to avoid them. For example, do not poke inside a toaster with a knife or fork.

6 **Project** Carry out a survey to find out what percentage of the adult population have experienced shocks and in what circumstances.

7 **Library a** Use resource books to find out about the first-aid treatment of electric shock victims.

b Use a biology book to find out more about the nerve system and the opposed muscle system of the human body.

21 WIRING PLUGS AND APPLIANCES

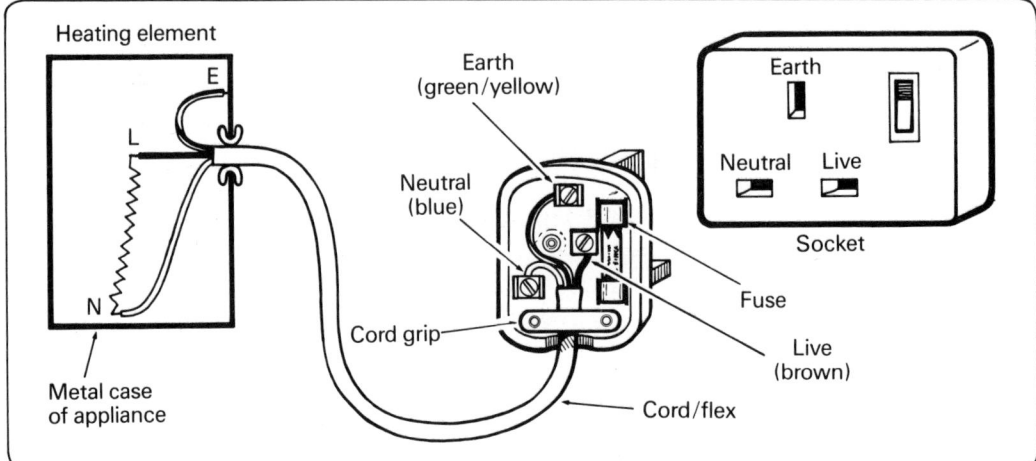

Electric appliances are connected to the mains by a three-pin plug. The plug is connected to the appliance by a flex containing usually three wires with their insulation colour-coded as follows:

LIVE – Brown
NEUTRAL– Blue
EARTH – Green and yellow stripes.

Inside the appliance, the earth lead is attached to the metal frame. It is therefore *vital that the green and yellow wire in the plug is connected to the earth pin and not the live!* If a mistake is made, the first person to handle the equipment may be killed. It could easily be a baby crawling across the floor and touching a fan heater.

Normally, current flows from live through the appliance (heating coil, motor, etc) and back to the local electricity substation by the neutral wire. If the live comes in contact with the metal frame, the current can flow back to the power station via the earth lead, then actually through the damp earth in the ground. At the power station the neutral line is connected to a large metal plate buried in the ground. Likewise at the house the local earth wire from the consumer unit must run down to earth.

The return of the 'fault' current via earth is shown in the diagram. If the earth connection is poor (having high resistance), the fault current may flow through the person touching the appliance instead and cause death.

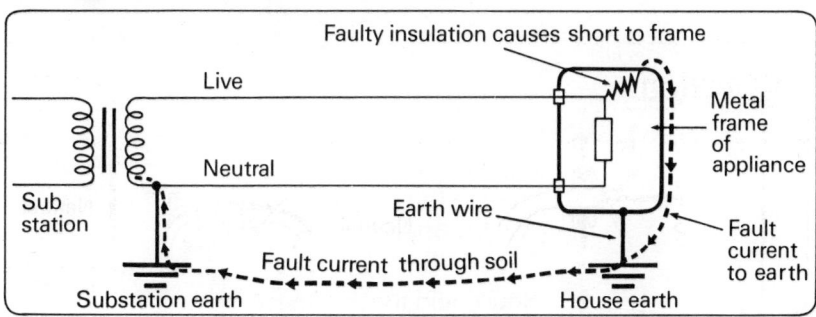

ACTIVITIES

Step 1   Wiring a plug

a   Practise wiring a plug; *then have it inspected.* Use this checklist:
  1   Is the colour code correctly observed?
  2   Is the cord grip tight?
  3   Will the cap fit on properly?
  4   Are all strands of the wire secured under the clamps?
  5   Is the fuse suitable for the appliance to be used?

b   Make a drawing to show how a plug is wired and state the colour code.

Step 2   Wiring an appliance

NOTE   This should be done as a teacher demonstration, or under close teacher supervision.

Take apart an appliance and find out how the wires are connected inside. Does the earth connect to the metal frame? What provisions has the designer made to stop the live wire coming in contact with the casing?

Step 3   Earth Test

It is important for there to be a complete circuit from the earth pin in a plug to the metal frame of the appliance.

a   We can check for continuity using a low voltage lamp and battery as shown overleaf. If the continuity is good, the lamp will light.

b Copy the diagram and carry out some earth tests.

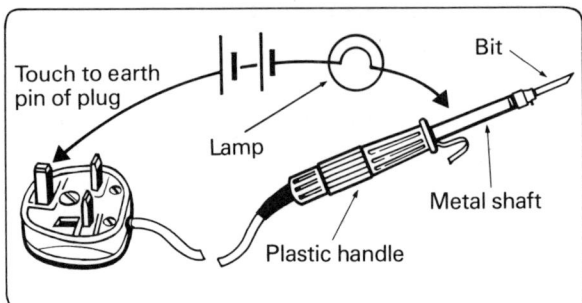

Questions

1 a The diagram shows an appliance with the switch fitted to the neutral lead. Will the appliance switch on and off?

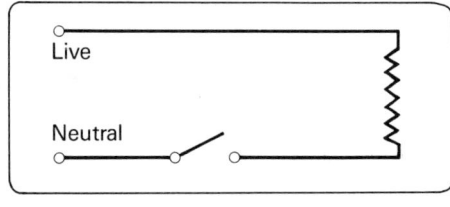

b How could the appliance be dangerous when switched off? Suppose it was a single bar electric fire?

2 Some appliances, such as razors, have no earth lead. They are double insulated. What could this mean?

3 a A fan heater contains a heater element, a fan, plus a switch. Copy the diagram and show how you would wire such a heater.

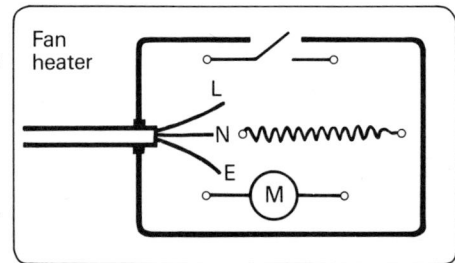

b How could a higher power setting be added?

4 Why should all appliances in a school, office or factory be given an earth continuity test at least once a year? What other inspection should be made?

5 The diagram represents the components of an electric iron. They include a switch, pilot lamp (mains voltage), heating element, and thermostat. Copy the diagram and use coloured pencils to show how the wires would be connected.

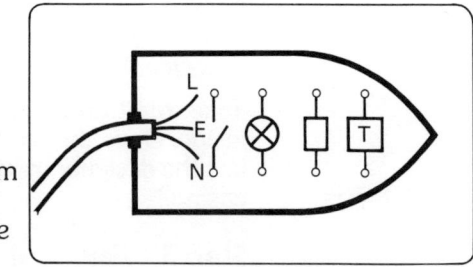

6 If a 5A fuse in the consumer unit kept blowing, would it be sensible to replace it by, say, a 30A one or a piece of copper wire?

7 Find out who is your school safety officer and what electrical surveys are carried out.

USEFUL DATA

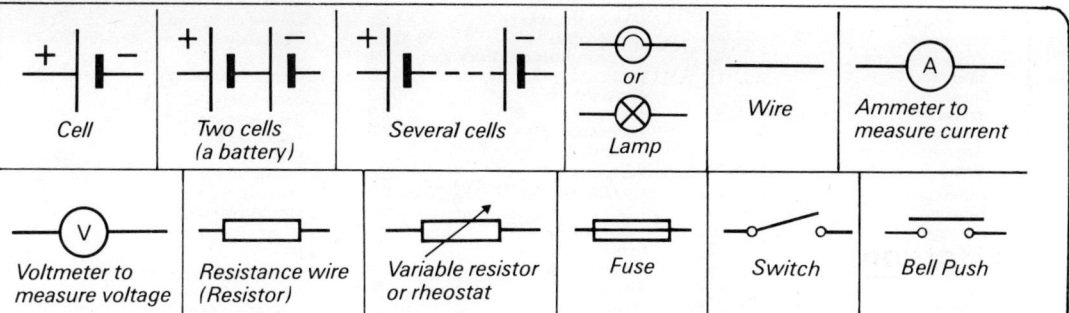

Quantity	Symbol	Unit	Abbreviation
Current	I	ampere	A
		milliampere	mA
Voltage, potential difference	V	volt	V
Resistance	R	ohm	Ω
		kilohm	$k\Omega$
		megohm	$M\Omega$
Power	P	watt	W
		kilowatt	kW
		megawatt	MW
Frequency	–	hertz	Hz
		kilohertz	kHz
		megahertz	MHz
Capacitance	C	farad	F
		microfarad	μF

Formulae

Ohm's Law	$V = IR \qquad R = \dfrac{V}{I} \qquad I = \dfrac{V}{R}$
Power	$P = IV \qquad P = \dfrac{V^2}{R} \qquad P = I^2 R$
Transformer	$\dfrac{V_{\text{SECONDARY}}}{V_{\text{PRIMARY}}} = \dfrac{N_{\text{SECONDARY}}}{N_{\text{PRIMARY}}}$
Resistors in series	$R = R_1 + R_2 + R_3$ etc.
Resistors in parallel	$\dfrac{1}{R} = \dfrac{1}{R_1} + \dfrac{1}{R_2}$ etc.

INDEX